Survivable Optical WDM Networks

OPTICAL NETWORKS SERIES

Series Editor
Biswanath Mukherjee, *University of California, Davis*

Other books in the series:

OPTICAL BURST SWITCHED NETWORKS
Jason P. Jue and Vinod M. Vokkarane, ISBN 0-387-23756-9

SURVIVABLE OPTICAL WDM NETWORKS

CANHUI (SAM) OU
SBC Services, Inc.

BISWANATH MUKHERJEE
University of California, Davis

Springer

Canhui (Sam) Ou
SBC Services, Inc.

Biswanath Mukherjee
University of California, Davis

Survivable Optical WDM Networks

Library of Congress Cataloging-in-Publication Data

A C.I.P. Catalogue record for this book is available
from the Library of Congress.

ISBN 978-1-4419-3749-0 e-ISBN 978-0-387-24499-0

Softcover reprint of the hardcover 1st edition 2005

9 8 7 6 5 4 3 2 1

springeronline.com

To our families and friends

Contents

List of Figures

Preface

The Topic

Optical-fiber technologies using wavelength-division multiplexing (WDM) are currently being researched and commercially deployed to satisfy our increasing bandwidth requirements because, by using WDM technologies, an optical fiber can support multiple non-overlapping wavelength channels, each of which typically operates at the data rate of 10 Gbps or 40 Gbps. In such a network, the failure of a network element, e.g., a fiber, can cause the failure of several wavelength channels, thereby leading to large data and revenue loss. The development of fault-management software, projected to grow significantly in the years ahead, is a top priority for both carriers and vendors. This book investigates the performance and design issues of survivable optical networks against failures.

The book first explores the problem of dynamic shared-path protection, which is desirably resource efficient because of backup sharing. It proves the NP-completeness of the problem, develops a heuristic to compute a feasible solution with high probability, and designs another heuristic to optimize resource consumption for a given solution.

As protection-switching time, resource efficiency, and scalability are primary concerns of a protection scheme, the book investigates from the network point of view sub-path protection, which achieves high scalability and fast recovery time for a modest sacrifice in resource utilization. The book then proceeds to explore from the connection point of view segment protection, which achieves fast recovery time and high resource efficiency.

While the transmission rate of a wavelength channel is high (typically STS-192 or STS-768), the bandwidth requirement of a typical connection request can vary from the full wavelength capacity down to STS-1 or lower. Different low-speed connections may request different bandwidth granularities as well as different protection schemes (dedicated, shared, or no protection). How to efficiently groom such low-speed connections while satisfying their protection requirements is investigated next. Both protection-at-lightpath level and protection-at-connection level are examined and evaluated.

Next-generation SONET/SDH technologies enable network operators to provide integrated data and voice services over their legacy SONET/SDH infrastructure to generate new revenue. An important open research problem on data over SONET/SDH (DoS) is survivability: SONET automatic protection switching is too resource inefficient for data services, and the protection mechanisms of data networks are too slow for mission-critical applications. The book proposes two approaches for provisioning survivable DoS connections. The approaches exploit the tradeoff between resource overbuild and fault-recovery time while utilizing the inverse-multiplexing capability of virtual concatenation to increase backup sharing.

Intended Audience

This book is intended to be a reference book on survivability of optical networks for industrial practitioners, researchers, and graduate students who work on and/or want to learn more about survivable optical networks and data-over-SONET/SDH networks.

The focus of the book is on the various alternative approaches for combatting failures, such as fiber cuts and switch failures, in mesh optical networks. Industrial practitioners and researchers should find the book to be a useful reference because it contains state-of-the-art techniques to address various design and operating issues on managing failures in mesh optical networks and data-over-SONET/SDH networks.

Organization of the Book

This book is divided into seven chapters. Chapter 1 presents an overview of optical communication networks and an overview of the book.

Chapters 2-4 focus on provisioning connections of full wavelength capacity, or lightpaths, in an optical WDM mesh network. These three chapters propose and investigate different approaches to explore the tradeoffs among resource efficiency (backup sharing), fault-recovery time, scalability, and implementation complexity.

Chapters 5-7 focus on provisioning connections of sub-wavelength granularity in an optical WDM mesh network. Chapter 5 presents survivable traffic grooming with dedicated protection. Chapter 6 discusses survivable traffic grooming with shared protection. Chapter 7 presents an overview of next-generation SONET/SDH technologies and investigates the survivability of data-over-SONET/SDH networks.

Feedback

We welcome emails from readers who wish to provide any sort of feedback: errors, comments, criticisms, and suggestions for improvements. Our contacts are as follows:

Canhui (Sam) Ou
sam.ou@sbc.com
http://networks.cs.ucdavis.edu/~ouc/

Biswanath Mukherjee
mukherje@cs.ucdavis.edu
http://networks.cs.ucdavis.edu/~mukherje/

CANHUI (SAM) OU
BISWANATH MUKHERJEE

Acknowledgments

Much of the book's material is based on research that we have conducted over the past couple of years with members of the Networks Research Laboratory at University of California, Davis, and we would like to acknowledge them as follows: Dr. Hui Zang, now at Sprint Advanced Technology Laboratories, for her collaboration on Chapters 2, 3, 5, and 6; Dr. Laxman H. Sahasrabuddhe, now at Park, Vaughan & Fleming LLP, for his collaboration on Chapters 2, 3, 5, 6, and 7; Dr. Keyao Zhu, now at Brion Technologies, for his collaboration on Chapters 3, 5, 6, and 7; Dr. Narendra Singhal, now at Microsoft Corp., for his collaboration on Chapters 3 and 7; Jing Zhang for her collaboration on Chapters 2 and 5; Hongyue Zhu for his collaboration on Chapter 5; Professor Charles Martel of the Computer Science Department at UC Davis for his collaboration on Chapter 7; and other members—Professor Dipak Ghosal, Professor Prasant Mohapatra, Dr. Glen Kramer (now at Teknovus), Smita Rai, Dr. Jian Wang (now at Florida International University), Dr. Wushao Wen (now at McAfee), Dr. Fred Xue, and Dr. Shun Yao (now at Park, Vaughan & Fleming LLP)—for their technical expertise and insightful discussion which have enabled us to better understand the subject matter.

A number of additional individuals whom we have the pleasure to collaborate with and whom we would like to acknowledge are the following: Dr. James Pan at Sprint Advanced Technology Laboratories, Robert MacDonald (retired from Sprint Advanced Technology Laboratories), Dr. Mike O'Brien (formerly with Sprint Advanced Technology Laboratories), and Dr. Ching-Fong Su at Fujitsu Laboratories of America.

This book wouldn't have been possible without the support of our research on survivable optical networks from several funding agencies as follows: US National Science Foundation (NSF) Grant Nos. ANI-9805285 and ANI-0207864; University of California Micro program; Alcatel Research & Innovation; and Sprint Advanced Technology Laboratories.

We would like to thank Steve Sposato and Raj Savoor at SBC for their support. We wish to thank Dr. Anpeng Huang at University of Waterloo for reviewing the manuscript. We gratefully acknowledge the people at Springer whom we interacted with—Alex Green and Melissa Guasch—for their encouragement and assistance.

Finally, we wish to thank our family members for their constant support and encouragement.

Chapter 1

INTRODUCTION

1.1 Optical Networking

With the explosive growth of the data network, optical-fiber technology employing wavelength-division multiplexing (WDM) [Mukherjee, 1997, Ramaswami and Sivarajan, 1998, Stern and Bala, 1999] has been researched as well as commercially deployed as the technology that can satisfy our increasing bandwidth requirements because of its potentially limitless capabilities, e.g., huge bandwidth and low signal attenuation [Mukherjee, 2000].

1.1.1 Telecommunication Networks

Figure 1.1 provides an overview of telecommunication networks, which consist of the access network, the metropolitan-area network, and the backbone network. Access networks, mainly in the form of trees, span a few kilometers to connect end customers to central offices. Widely-used technologies for access networks are dial-up, digital subscriber line (DSL), cable modem, and others. Metropolitan-area networks cover a few 10s to a few 100s of kilometers, and employ mainly the SONET/SDH ring infrastructure. Backbone networks can span a few 100s to a few 1000s of kilometers. Historically, backbone networks have been interconnections of stacked rings. Due to the inefficiency and poor scalability of interconnected rings, backbone networks are expected to migrate to the resource-efficient and scalable meshes.

Backbone networks can be either packet switched or circuit switched. While optical packet-switched networks are somewhat futuristic, wavelength-routed (i.e., circuit-switched) networks are becoming realistic with the maturity of optical crossconnects (OXCs) [Cisco white paper, phtm]. In this book, we focus on wavelength-routed WDM mesh networks.

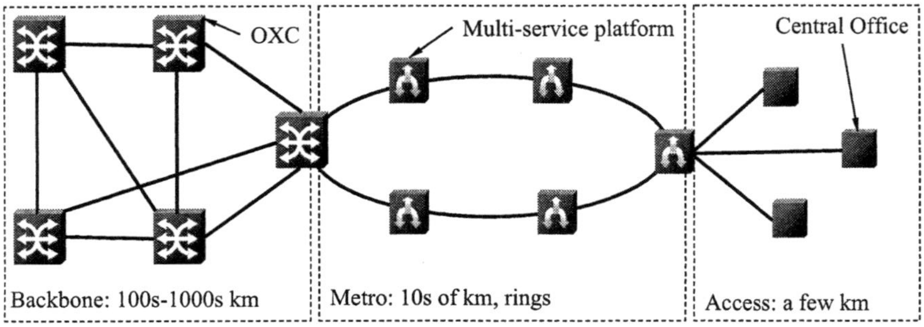

Figure 1.1. Overview of telecommunication networks.

1.1.2 Wavelength-Routed WDM Mesh Networks

In a wavelength-routed network, *lightpath* is the basic communication approach [Chlamtac et al., 1992]. A lightpath is a direct optical transmission channel between two network nodes, and it can operate at the rate of a few gigabits per second (Gbps). An example of a lightpath is a circuit between two client network elements (NE), e.g., IP routers, ATM switches, etc., as in Fig. 1.2. In the absence of any wavelength-conversion device, a lightpath is required to occupy the same wavelength throughout its path in the network; this is referred to as the *wavelength-continuity* property of a lightpath. This requirement can be relaxed if the network has wavelength converters (e.g., referred to as semilightpath in [Chlamtac et al., 1996]). The problem of routing lightpaths in a WDM network is known as the routing and wavelength-assignment (RWA) problem and it has been extensively studied [Chlamtac et al., 1992, Chlamtac et al., 1996, Ramaswami and Sivarajan, 1995, Banerjee and Mukherjee, 1996, Chen and Banerjee, 1996, Zang et al., 2000].

An optical network with optical crossconnects is expected to have dynamic lightpath-provisioning capability. Figure 1.2 provides a logical view of such a network. Conceptually, it has transport, control, and management planes. The transport plane provides high-speed transmission; the control plane provides real-time signaling; and the management plane manages the network resources, monitors network state, and provides billing (and other) capabilities. Inside an administrative domain, OXCs are interconnected through fibers in a mesh topology, and they communicate with one another through the Inter-

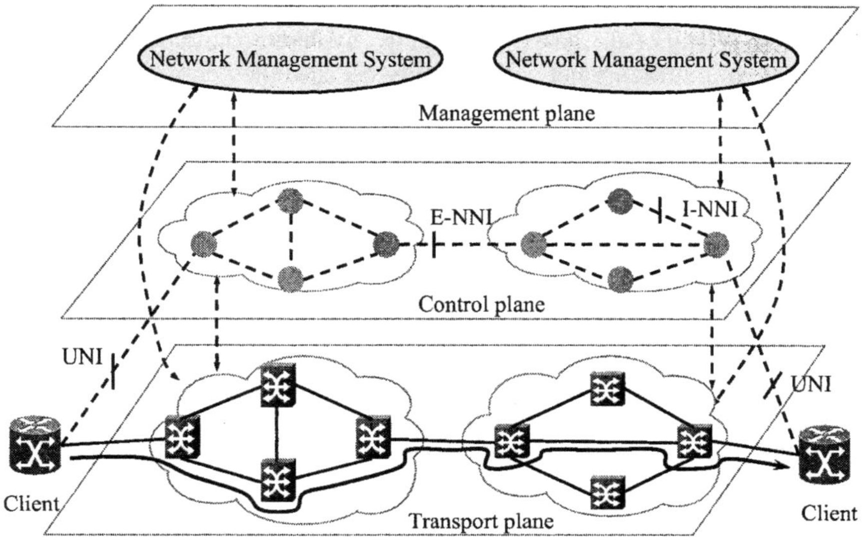

Figure 1.2. Logical view of a backbone network.

nal Network Network Interface (I-NNI). The communication between different administrative domains is through the External Network Network Interface (E-NNI).

A lightpath-provisioning request can be initiated by the network operator from a network management system (NMS), or by a client network element through the standard User Network Interface (UNI) [Optical Internetworking Forum (OIF), 2004]. Upon the arrival of such a request, the route-computation module (which can be part of NMS or part of the control plane) computes a route based on the current network state. If an eligible route is found, the control plane will signal the OXCs along the route to configure themselves.

1.1.3 Survivable WDM Mesh Networks

In a wavelength-routed WDM network, the failure of a network element (e.g., port on a client equipment, optical-layer hardware, fiber link, crossconnect, etc.) can cause the failure of several lightpaths, thereby leading to large data and revenue loss. Fault-recovery mechanisms, e.g., protection and restoration, are essential for a network to survive from such failures. If backup resources (routes and wavelengths) are pre-computed and reserved in advance, we call it a *protection* scheme [Ramamurthy et al., 2003, Doshi et al., 1999, Mohan et al., 2001, Ellinas et al., 2003, Fumagalli et al., 2003, Goyal et al., 2002, Koo and Subramaniam, 2002]. Otherwise, when a failure occurs, if another route and a free wavelength have to be discovered dynamically for each disrupted connection, we call it a *restoration* scheme [Ramamurthy and Mukherjee, 1999, Wang et al., 2002c, Wang et al., 2002b]. Generally, dynamic restoration schemes are more efficient in utilizing network capacity because they do not allocate spare capacity in advance, and they provide resilience against different kinds of failures (including multiple failures); but protection schemes have faster recovery time and they can guarantee recovery from disrupted services they are designed to protect against (a guarantee which restoration schemes cannot provide).

Protection schemes can be classified as ring protection and mesh protection. Ring-protection schemes include Automatic Protection Switching (APS) and Self-Healing Rings (SHR). Both ring protection and mesh protection can be further divided into two groups: path protection and link protection. In *path protection*, traffic is rerouted through a link-disjoint backup route (*backup path*) once a link failure occurs on its working path (*primary path*).[1] In *link protection*, traffic is rerouted only around the failed link. While path protection leads to efficient utilization of backup resource and lower end-to-end propagation delay for the recovered route, link protection provides faster protection-switching time. Recently, researchers have proposed *sub-path protection* by dividing a primary path into a sequence of segments and protecting each segment separately [Ou et al., 2002, Ho and Mouftah, 2002, Anand et al., 2002, Xu et al., 2003a, Wang et al., 2004]. Compared with path protection, sub-path protection can achieve high scalability and fast recovery for a modest sacrifice in resource utilization.

Link, sub-path, and path protection schemes can be dedicated or shared [Doshi et al., 1999, Lumetta et al., 2000]. In *dedicated protection*, there is no sharing between backup resources, while in *shared protection*, backup wavelengths can be shared on some links as long as their protected segments (links, sub-paths, paths) are mutually diverse.[2] OXCs on backup paths cannot be configured until the failure occurs if shared protection is used. So, recovery time in shared protection is longer but its resource utilization is better than dedicated protection.

---

[1] Single-fiber failures are the dominant failures in communication networks, and they occur mainly due to construction equipment ploughing through buried cable. Node failures are relatively rare because switch fabric and switch-control unit in a carrier-class node are typically dedicated (1+1) – i.e., master/slave – protected. A node's port cards, however, are generally not 1+1 protected since they take up the bulk of the space (perhaps over 80%) and cost of a node (switch); also a port-card failure can be handled as link and/or wavelength channel failure(s). Therefore, nodes are more robust than links. Nevertheless, node failures can be accommodated by calculating node-disjoint routes. And node failures are important to protect against in scenarios where an entire node (or a collection of nodes in a part of the network) may be taken down, possibly due to a natural disaster or by a malicious attacker.

[2] The resources on a backup path can be shared among different connections as long as these connections are not in the same shared-risk group (SRG). Since link failure is the dominant failure scenario and fibers are laid in bundles so that multiple fibers may break when there is a bundle cut, shared-risk link group (SRLG) is a commonly-used method for diversely routing various primary and backup paths of connections [Strand et al., 2001].

Protection in optical WDM networks is an extremely important problem and is the focus of this book.

1.2 An Overview of the Book

1.2.1 Shared-Path Protection

For dynamic survivable lightpath provisioning in optical WDM mesh networks, we prove that the problem of finding an eligible pair of working and backup paths for a new lightpath request requiring shared-path protection under the current network state is NP-complete. Then, we develop a heuristic, called CAFES, to compute a feasible solution with high probability and another heuristic, called OPT, to optimize resource consumption for a given solution. Shared-path protection will be presented in Chapter 2.

1.2.2 Sub-Path Protection for Scalability and Fast Recovery

We propose sub-path protection, which is a generalization of shared-path protection. The main ideas of sub-path protection are (1) to partition a large optical network into smaller domains, and (2) to apply shared-path protection to the optical network such that an intra-domain lightpath does not use resources of other domains and the primary/backup paths of an inter-domain lightpath exit a domain (and enter another domain) through a common domain-border node. We mathematically formulate the routing and wavelength-assignment (RWA) problem under sub-path protection for a given set of lightpath requests, prove that the problem is NP-complete, and develop a heuristic to find efficient solutions. Comparisons between sub-path protection and shared-path protection on a nationwide network with dozens of wavelengths per fiber show that, for a modest sacrifice in resource utilization, sub-path protection achieves improved survivability, much higher scalability, and significantly reduced fault-recovery time. Sub-path protection will be presented in Chapter 3.

1.2.3 Segment Protection

We unify various forms of segment protection into generalized segment protection (GSP). In GSP, the working path of a lightpath is divided into multiple overlapping working segments, each of which is protected by a node/link disjoint backup segment. We design an efficient heuristic which, upon the arrival of a lightpath request, dynamically divides a judiciously-selected working path into multiple overlapping working segments and computes a backup segment for each working segment while accommodating backup sharing. Compared to the widely-considered share-path protection scheme, GSP achieves much lower blocking probability and shorter protection-switching time for a small sacrifice in control and management overhead.

Based on generalized segment protection, we present a new approach to provisioning lightpath requests according to their differentiated quality-of-protection (QoP) requirements. We focus on one of the most important QoP parameters—namely, protection-switching time—since lightpath requests may have differentiated protection-switching-time requirements. For example, lightpaths carrying voice traffic may require 50-ms protection-switching time while lightpaths carrying data traffic may have a wide range of protection-switching-time requirements. Numerical results show that our approach achieves significant performance gain which leads to a remarkable reduction in blocking probability. Segment protection will be presented in Chapter 4.

1.2.4 Survivable Traffic Grooming–Dedicated Protection

A typical connection request may require bandwidth less than that of a wavelength, and it may also require protection from network failures, typically fiber cuts. Based on a generic grooming-node architecture, we propose two approaches—protection-at-lightpath (*PAL*) level and protection-at-connection (*PAC*) level—for grooming a connection request. For dedicated protection, we prove that the problem of provisioning a connection under PAC is $\mathcal{NP}$-complete, propose effective heuristics for both schemes, and define comprehensive performance metrics to compare PAL with PAC with respect to wavelength/grooming-port efficiency. Our findings are as follows. Under today's typical connection-bandwidth distribution where lower bandwidth connections outnumber higher bandwidth connections, PAC outperforms PAL (in terms of bandwidth-blocking ratio, lightpath utilization, and wavelength utilization) if the number of grooming ports is large; however, PAL outperforms PAC (in terms of bandwidth-blocking ratio and grooming-port utilization) when the number of grooming ports is moderate or small. Survivable traffic grooming with dedicated protection will be presented in Chapter 5.

1.2.5 Survivable Traffic Grooming–Shared Protection

We propose three approaches—protection-at-lightpath (*PAL*) level, mixed protection-at-connection (*MPAC*) level, and separate protection-at-connection (*SPAC*) level—for grooming a connection request with shared protection. These three schemes explore different ways of backup sharing, and they trade off between wavelengths and grooming ports. Since the existence version of the problem for provisioning one connection request with shared protection is $\mathcal{NP}$-complete, we propose effective heuristics. Our findings are as follows. Under today's typical connection-bandwidth distribution where lower bandwidth connections outnumber higher bandwidth connections, 1) it is beneficial to groom working paths and backup paths separately, as in PAL and SPAC; 2) separately protecting each individual connection—i.e., SPAC—yields the best performance when the number of grooming ports is sufficient; and 3) protecting each specific lightpath—i.e., PAL—achieves the best performance when the number of grooming ports is moderate or small. Survivable traffic grooming with shared protection will be presented in Chapter 6.

1.2.6 Survivable Virtual Concatenation for Data over SONET/SDH

An important open research problem on data over SONET/SDH (DoS) is survivability: SONET automatic protection switching is too resource inefficient for data services, and the protection mechanisms of data networks are too slow for mission-critical applications. We propose two approaches, PIVM and PREV, for provisioning survivable DoS connections. Our approaches exploit the tradeoff between resource overbuild and fault-recovery time while utilizing the inverse-multiplexing capability of virtual concatenation to increase backup sharing. Our results show that PIVM achieves low resource overbuild and much faster fault recovery than that of data networks, and PREV achieves fast fault recovery comparable to SONET 50-ms protection (for typical US backbone networks) while still achieving modest backup sharing. We further investigate the tradeoff between network blocking performance and network control and management complexity resulting from the number of paths M a connection can be inversely multiplexed onto: larger M leads to more freedom in routing and better network performance but increases network control and management complexity. Our results indicate that the network blocking performance for small values of M (e.g., $M = 2$ for some topologies) is almost as good as the case in which M is infinity. Survivable virtual concatenation will be presented in Chapter 7.

Chapter 2

SHARED-PATH PROTECTION FOR RESOURCE EFFICIENCY

2.1 Introduction

We consider the problem of dynamic survivable lightpath provisioning against single-fiber failures. Specifically, we focus on shared-path protection because of its desirable resource efficiency resulting from backup sharing. Protection approaches to optimizing resource utilization for a given traffic matrix [Crochat et al., 2000, Modiano and Narula-Tam, 2002, Ramamurthy et al., 2003, Van Caenegem et al., 1998] do not apply because lightpath requests come and go in the dynamic provisioning case. Under such a scenario, a network management system needs to compute two link-disjoint paths—a dedicated working path and a shared backup path—for an incoming lightpath request based on the current network state. We concentrate on computing link-disjoint paths for each incoming lightpath request with the assumptions that existing lightpaths cannot be disturbed and no knowledge of future arrivals is available at the time of provisioning this lightpath request. While we consider full wavelength-convertible networks here, the extension to the wavelength-continuous case is straightforward.

Much work has been conducted on dynamic survivable lightpath protection in an optical WDM network and on dynamic routing of restorable bandwidth-guaranteed connections in a Multi-Protocol Label Switching (MPLS) network. Although some papers are devoted to the MPLS context, their basic ideas—with appropriate variations, e.g., quantized bandwidth granularities—are applicable to the shared-path protection problem in a WDM mesh network with full wavelength conversion at each node. Table 2.1 provides an overview of some related work, more elaboration on which follows. In what follows, the term "lightpath" will be used in WDM context while the term "connection" will be used in MPLS context.

The most desirable property of shared-path protection is its resource efficiency resulting from backup sharing. Consequently, how to increase backup sharing based on different cost models is of particular interest and has been reported in [Bouillet et al., 2002b, Elie-Dit-Cosaque et al., 2002, Kodialam and Lakshman, 2000, Li et al., 2002, Liu et al., 2001, Su and Su, 2001a, Su and Su, 2001b]. Since backup sharing depends on the routes of working paths, most of existing work computes a backup path after the working path is determined. A natural questions is: Can we jointly compute a working path and a backup path for a lightpath request in polynomial time in a way similar to Suurballe's algorithm[1]? We show that the answer is no by proving that the problem of finding an eligible pair of working and backup paths under shared-path-protection constraints for a lightpath request with respect to existing lightpaths is NP-complete. Then, we develop a heuristic, called CAFES, to compute a feasible solution (i.e., two link-disjoint paths) with high probability. A drawback of computing a backup path after fixing the working path is that the working and backup paths combined may use more resources than necessary. Therefore, we provide a new heuristic, called OPT, to iteratively optimize the resource consumption for a given solution. While our focus is on dynamic lightpath provisioning, the two heuristics can be readily applied to static lightpath provisioning in which the traffic matrix is known a priori. For example, we can apply CAFES to find a feasible solution for every lightpath request in the traffic matrix in the first phase; and in the second phase, we can apply OPT to every solution found in the first phase to optimize the overall resource consumption.

While it is desirable to have complete information about the routing and wavelength assignment of the existing lightpaths to decide backup sharing, complete information may not always be available due to control and management concerns. Three scenarios—complete, partial, and no information—have been introduced in [Kodialam and Lakshman, 2000] and further examined in [Kodialam and Lakshman, 2001, Qiao and Xu, 2002] to quantify the impact of the amount of available information on backup sharing. The amount of information in the complete-information scenario can be reduced without sacrificing backup sharing by aggregating backup-sharing information via various techniques as shown in [Elie-Dit-Cosaque et al., 2002, Liu et al., 2001, Mohan et al., 2001, Su and Su, 2001a, Su and Su, 2001b]. Our study utilizes aggregated information, which will be discussed in Section 2.2.1.

One possible limitation of shared-path protection is that backup paths may sometimes become longer due to backup sharing [Ramamurthy et al., 2001]. The relation between backup sharing and backup-path hop distance has been shown to be such that one trades off another in [Bouillet et al., 2002b, Xiong

[1] Suurballe's algorithm [Suurballe and Tarjan, 1984] jointly computes two link-disjoint paths of minimal total cost without consideration of backup sharing.

et al., 2003]. Our study prevents backup paths from being unnecessarily detoured while encouraging backup sharing in a way related to the approach in [Xiong et al., 2003].

While the dynamic shared-path-protection problem can be formulated as an integer linear program (ILP) [Qiao and Xu, 2002, Su and Su, 2001b, Xiong et al., 2003], ILPs are not scalable based on current computational power. As reported in [Xiong et al., 2003], the average processing time per connection request is 0.5 second for a 15-node network with 200 requests (on a PC with a 1.5 GHz CPU); the average processing time per connection request quickly increases to 8.5 seconds for a 46-node network with 200 requests. Therefore, we resort to efficient heuristics in this study.

The rest of this chapter is organized as follows. Section 2.2 formally states the problem and analyzes its complexity. Section 2.3 presents the CAFES heuristic for finding a feasible solution. Section 2.4 presents another heuristic, OPT, for optimizing a given solution. Section 2.5 evaluates the performance of our heuristics via simulations. Section 2.6 concludes this chapter.

2.2 Problem Statement and Complexity Analysis

2.2.1 Problem Statement

We first define the notations and then formally state the dynamic shared-path-protected lightpath-provisioning problem. A network is represented as a weighted, directed graph $G = (V, E, C, \lambda)$, where V is the set of nodes, E is the set of unidirectional fibers (referred to as links), $C : E \rightarrow R^+$ is the cost function for each link (where R^+ denotes the set of positive real numbers), and $\lambda : E \rightarrow Z^+$ specifies the number of wavelengths on each link (where Z^+ denotes the set of positive integers).

We use λ_f^e to denote the number of free wavelengths on link $e \in E$. We denote the set of existing lightpaths by $\mathcal{L} = \{(l_w^i, l_b^i, t_a^i, t_h^i)\}$, where the quadruple $(l_w^i, l_b^i, t_a^i, t_h^i)$ specifies the working path, the backup path, the arrival time, and the holding time, in order, for the ith lightpath. We denote the current lightpath request by (l_w, l_b, t_a, t_h). We represent the cost of l_w and l_b using $C_w(l_w)$ and $C_b(l_w, l_b)$, respectively.

We associate a conflict set with a link[2] to identify the sharing potential between backup paths. The conflict set ν_e for link e defines the set of links used by those working paths whose backup paths utilize wavelengths on link e. The conflict set ν_e for link e can be represented as an integer set, $\{\nu_e^{e'} \mid \forall e' \in$

[2]In the wavelength-continuous case, we would associate a conflict set to a wavelength. The conflict set defined here is similar to the conflict vector in [Mohan et al., 2001], the aggregated square matrix in [Liu et al., 2001], and the "bucket" link metric in [Su and Su, 2001b], but it is more general in the sense that the conflict set can model wavelength-continuous networks, wavelength-convertible networks, and networks of sparse wavelength-conversion capability [Iness and Mukherjee, 1999].

Table 2.1. Comparison of related work on dynamic shared-path-protected lightpath/connection provisioning on WDM/MPLS mesh networks.*

Research Work	Objective	Path Link	Cen. Dis.	Info.	λ Con.	Det. Pro.	Contributions (in brief)
[Bouillet et al., 2002a] [Bouillet et al., 2002b]	Minimize the total cost of working and backup paths for each lightpath	P	C	A	Y	P D	Stochastic approaches; cost model; K-shortest path routing.
[Elie-Dit-Cosaque et al., 2002]		P	D	F, P	Y	D	Protection-sharing table.
[Mohan et al., 2001]		P	C	A	N	D	Primary-backup sharing; Cost model for route computation.
[Ramamurthy et al., 2001]		P	D	F	Y	D	Performance comparison of different schemes.
[Su and Su, 2001a] [Su and Su, 2001b]		L	D	F	Y	D	Bucket-based link metric; ILP & two-step heuristic.
[Xin et al., 2001b]		P	C	F	Y	D	K-shortest path routing.
[Xiong et al., 2003]		P	C & D	F & P	Y	D	ILP formulations.
Our work		P	C	A	Y	D	NP-complete proof; heuristic for optimization; heuristic for finding a feasible solution.
[Kodialam and Lakshman, 2000] [Kodialam and Lakshman, 2001]	Minimize the total cost of working and backup paths for each connection	P & L	C & D	F, N, P	Y	D	ILPs for different scenarios & a heuristic based on primal-dual and LP-relaxation.
[Li et al., 2002]		P	C	F	Y	D	Two-step heuristic using a bucket-like link metric; distributed signaling.
[Liu et al., 2001]		P	D	A	Y	D	Aggregating per-flow information with a matrix; successively updating existing backups.
[Qiao and Xu, 2002]		P	D	P	Y	D	ILP & two-step heuristic; distributed signaling.

Path/Link: Path protection or link protection.
Cen./Dis.: Whether the algorithm is centralized or distributed.
Info.: The amount of information needed. A: aggregated lightpath/connection information; F: full per-lightpath/connection information;
N: no information about existing lightpaths/connections; P: partial information about existing lightpaths/connections.
λ Con.: Whether the work applies to wavelength-continous network or wavelength-convertible network.

* Please note that, while the works in [Kodialam and Lakshman, 2000, Kodialam and Lakshman, 2001, Li et al., 2002, Liu et al., 2001, Qiao and Xu, 2002] are devoted to MPLS networks, their basic ideas are applicable to WDM networks.

$E, 0 \leq \nu_e^{e'} \leq \lambda(e')\}$, where $\nu_e^{e'}$ specifies the number of working paths that traverse link e' and are protected by link e (and their corresponding backup paths traverse link e). The number of wavelengths reserved for backup paths on link e is thus $\nu_e^* = \max_{\forall e'}\{\nu_e^{e'}\}$. Clearly, the union of the conflict sets for all the links aggregates the per-lightpath-based information, and the size of the conflict set depends only on the number of links, not on the number of lightpaths. In the absence of such a mechanism as conflict set, per-lightpath-based information is necessary for identifying shareable backup channels [Bouillet et al., 2002a]. It is, thus, advantageous to use conflict set since the number of lightpaths can be significantly more than the number of links.

The working and backup paths l_w and l_b satisfy the *shared-path-protection constraints* with respect to the existing lightpaths as follows:

C.1 l_w and l_b are link disjoint[3].

C.2 l_w and l_w^i, $1 \leq i \leq |\mathcal{L}|$, do not utilize the same wavelength on any common link they traverse.

C.3 l_w does not share any wavelength with l_b^i, $1 \leq i \leq |\mathcal{L}|$, on any common link they traverse.

C.4 l_b and l_b^i can share a wavelength on a common link only if l_w and l_w^i are link disjoint.

We now formally state the dynamic shared-path-protected lightpath-provisioning problem as follows: Given a WDM network as $G = (V, E, C, \lambda)$ and the set of existing lightpaths (or the associated conflict sets $\{\nu_e \mid e \in E\}$), route each incoming lightpath request under shared-path-protection constraints while minimizing the total cost of the working and backup paths. In the following subsection, we show that the existence version of this problem is NP-complete.

2.2.2 Complexity Analysis

We formally state the decision version of the dynamic shared-path-protected lightpath-provisioning (DSPLP) problem below and prove that it is NP-complete.

Instance: A graph $G = (V, E, C, \lambda)$, the set of existing lightpaths $\mathcal{L}$ (or the set of conflict sets $\{\nu_e \mid e \in E\}$), and a lightpath request from s to d ($s, d \in V$).

Question: Do there exist from s to d two paths, l_w and l_b, such that they satisfy the shared-path-protection constraints with respect to the existing lightpaths?

[3]The working and backup paths should be link disjoint to protect against link failures and they should also be node disjoint to protect against node failures. For this study, we shall focus on link disjointedness since link failures (fiber cuts) are the predominant form of failures in telecom networks, while several forms of node failures can be combatted through 1+1 redundancy in the nodal hardware components.

THEOREM 2.1 *DSPLP is NP-complete.*

Proof: *Please refer to Appendix 2.A.* ■

2.3 Compute A FEasible Solution (CAFES)

As the existence version of the problem is NP-complete, we resort to heuristics. In this section, we design a backtracking-based heuristic, called CAFES, to compute an eligible pair of working and backup paths for a lightpath request. In Section 2.4, we develop a general optimization procedure, called OPT, to iteratively optimize the resource consumption of the working and backup paths for a given solution (i.e., two link-disjoint paths).

A widely-used approach for computing a feasible solution is the so-called two-step approach, which first computes a least-cost path as the working path and then computes as the backup path a link (or node) disjoint path of least additional cost. A limitation of the two-step approach is that it cannot find a solution in a trap topology [Dunn et al., 1994], which is elaborated below, even though a solution exists. An improvement is to compute K working/backup path pairs (typically by applying a K-shortest-path algorithm to compute K candidate working paths and computing a backup path for each candidate working path) and select the pair of minimal cost. As the candidate working paths of the K shortest path pairs may share some common links, enumerating paths with practical values of K, e.g., 2 or 3, may be susceptible to a trap topology as well. Meanwhile, a non-trap topology can later become a trap topology after all the wavelengths on some links are used up. Furthermore, due to backup sharing, trap situations can arise even though the topology is not a trap topology. A possible drawback of enumerating paths is the lack of backtracking, i.e., the information gathered from enumerating the first i paths is not utilized in enumerating the $(i+1)^{th}$ path. We analyze the characteristics of two types of trap situations—trap topology and backup-sharing-caused trap—and propose a backtracking-based solution.

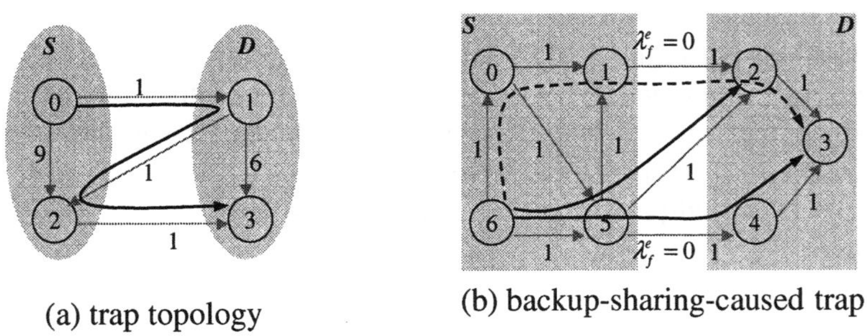

Figure 2.1. Trap situations. Solid gray lines represent links; solid black lines denote working paths; dashed black lines denote backup paths; and the number besides a link represents the cost of that link.

2.3.1 Trap Topology

For the example network in Fig. 2.1a, the two-step approach cannot find two link-disjoint paths from node 0 to node 3 (even though they exist) because the graph is disconnected after the removal of the working path (which is $\langle 0, 1, 2, 3 \rangle$).

We introduce backtracking based on network flow to overcome the trap situation. Let S be the set of nodes reachable from the source node after removing the links which are not link disjoint to the working path. Let D be the complement of S. (S, D) is referred to as a *cut*. We refer to a link as a *backhaul* link with respect to cut (S, D) if its source node is in D and its destination node is in S. For example, link $\langle 1, 2 \rangle$ in Fig. 2.1a is a backhaul link.

From the viewpoint of network flow, the objective of route computation is to push two units of link-disjoint flow from node s to node d. When a backup path cannot be found, it implies that no flow link-disjoint to the working path can be pushed from set S to set D. If the working path does not use any backhaul link, then clearly there is no link-disjoint flow from node s to node d. However, if the working path does use backhaul links, then it needs to traverse multiple links whose upstream nodes are in set S and downstream nodes are in set D. For example, in Fig. 2.1a, the working path traverses two such links $\langle 0, 1 \rangle$ and $\langle 2, 3 \rangle$. The backup path cannot use those links due to the link-disjoint constraint. As a result, no link-disjoint flow can be pushed from set S to set D and no link-disjoint backup path can be found.

In a two-step approach, working paths can traverse backhaul links because the two-step approach is greedy in the sense that it always selects a least-cost path as the working path. Meanwhile, due to the lack of backtracking in computing the backup path, a two-step approach cannot avoid backhaul links intelligently and, as a result, some lightpath requests may get blocked.

In our proposed approach CAFES, if a backup path is not found, it identifies the set of backhaul links, increases the cost of the backhaul links to some large value such as the sum of the costs of all the links in the network, and restarts the two-step process. This way, the working path will avoid, if possible, these backhaul links, and the backup path will have a chance to reach nodes in D. For example, if we artificially increase the cost of the backhaul link $\langle 1, 2 \rangle$ to 1000 in Fig. 2.1a and recompute the working path, which turns out to be $\langle 0, 1, 3 \rangle$, we are able to compute a link-disjoint backup path $\langle 0, 2, 3 \rangle$.

2.3.2 Backup-Sharing-Caused Trap

Another form of trap situation is termed backup-sharing-caused trap, which is illustrated by using the following example. Consider the network state in Fig. 2.1b. One existing lightpath with working path $\langle 6, 5, 4, 3 \rangle$ and backup path $\langle 6, 0, 1, 2, 3 \rangle$ is shown (other existing lightpaths are not shown). Suppose, to protect one more working path traversing link $\langle 6,5\rangle$, links $\langle 5,4\rangle$ and $\langle 1,2\rangle$ both need to allocate one more free wavelength (i.e., $\nu_{\langle 5,4\rangle}^{\langle 6,5\rangle} = \nu_{\langle 5,4\rangle}^{*}$ and $\nu_{\langle 1,2\rangle}^{\langle 6,5\rangle} = \nu_{\langle 1,2\rangle}^{*}$) and these two links have no free wavelength. Assume other links have free wavelengths and the cost of each link is unity. When a new lightpath request from node 6 to node 2 arrives, a two-step approach may compute path $\langle 6,5,2\rangle$ as the working path. As a result, no backup can be found because no more link-disjoint flow can be pushed from S to D (please note: $\langle 6,5,2\rangle$ and $\langle 6,5,4,3\rangle$ are not link disjoint; $\nu_{\langle 5,4\rangle}^{\langle 6,5\rangle} = \nu_{\langle 5,4\rangle}^{*}$; $\lambda_f^{\langle 5,4\rangle} = 0$; $\nu_{\langle 1,2\rangle}^{\langle 6,5\rangle} = \nu_{\langle 1,2\rangle}^{*}$; and $\lambda_f^{\langle 1,2\rangle} = 0$).

Again, we use backtracking to overcome this situation. Define a link $\langle m,n\rangle$ as a *conflicting* link with respect to cut (S,D) if there exists link $\langle p,q\rangle$ ($p \in S, q \in D$) such that $\nu_{\langle p,q\rangle}^{\langle m,n\rangle} = \nu_{\langle p,q\rangle}^{*}$ and $\lambda_f^{\langle p,q\rangle} = 0$. For example, link $\langle 6,5\rangle$ in Fig. 2.1b is a conflicting link as $\nu_{\langle 5,4\rangle}^{\langle 6,5\rangle} = \nu_{\langle 5,4\rangle}^{*}$ and $\lambda_f^{\langle 5,4\rangle} = 0$.

If the second minimal-cost path is not found and conflicting links exist, CAFES increases the cost of the conflicting links to some large value and restarts the two-step process. For example, if we artificially increase the cost of the conflicting link $\langle 6,5\rangle$ to 1000 in Fig. 2.1b and recompute the first minimal-cost (working) path, which turns out to be $\langle 6,0,5,2\rangle$, we are able to compute a link-disjoint minimal-cost backup path $\langle 6,5,1,2\rangle$ as it can share the wavelength-link $\langle 1,2\rangle$ with the existing backup $\langle 6,0,1,2,3\rangle$.

If there exist chained trap situations, in which some traps do not appear until some others are processed, we can recursively apply this procedure. We introduce a parameter k to limit the number of recursions. The parameter k can be considered as the maximum number of trap situations we want to process.

A formal specification of our heuristic, CAFES, is in Algorithm 2.1. In the algorithm, ϵ is a small number, e.g., 10^{-4}. The backup cost function C_1 is used to meet the shared-path-protection constraints C.1-C.4 (the first and last cases in C_1's definition) and to increase backup sharing (the second case). The last case of C_1's definition, $C(e)+\epsilon\cdot(\lambda(e)-\lambda_f^e)\cdot C(e)$, is used for load balancing: when there are two eligible backup paths of the same cost, the less loaded path will be chosen as backup. (Recall that $C(e)$ is the cost of link e.)

The computational complexity of CAFES is $O(k \times |E|^2)$. In particular, the complexities of Steps 1-6 are $O(1)$, $O(|V|^2)$, $O(|E|^2)$, $O(|E|^2)$, $O(|E|)$, and $O(1)$, respectively; Steps 2-6 repeat for at most $k+1$ times.

2.4 Optimization (OPT)

Given a feasible solution l_w and l_b for a lightpath from node s to node d, we develop a heuristic, called OPT, to minimize the total cost of l_w and l_b, $C_w(l_w) + C_b(l_w, l_b)$. Similar to Algorithm L in [Doshi et al., 1999], OPT iteratively refines l_w and l_b. In one iteration, OPT first recomputes l_w with l_b

Algorithm 2.1 CAFES

Input: $G = (V, E, C, \lambda)$, $\nu = \{\nu_e \mid e \in E\}$, $s, d \in V$, k
Output: Two paths l_w and l_b satisfying constraints C.1-C.4, or NULL if no such paths are found.

1 $l_w' \leftarrow$ NULL,

2 compute a minimal-cost path l_w on G from node s to node d; return NULL if l_w is not found or if $l_w' = l_w$,

3 compute a minimal-cost path l_b from node s to node d using cost function:

$$C_1(e) := \begin{cases} +\infty & \text{if } e \in l_w \vee (\lambda_f^e = 0 \wedge \\ & (\exists e' \in l_w, \nu_e^{e'} = \nu_e^*)) \\ \epsilon \times C(e) & \text{if } \forall e' \in l_w, \nu_e^{e'} < \nu_e^* \\ C(e) + \epsilon \cdot (\lambda(e) - \lambda_f^e) \cdot C(e) & \text{otherwise} \end{cases}$$

return $\langle l_w, l_b \rangle$ if l_b is found; return NULL if l_b is not found and $k = 0$,

4 compute the set of backhaul links L_b and the set of conflicting links L_c,

5 increase the cost of any link in L_b and L_c to some large value, and

6 $k \leftarrow k - 1$, $l_w' \leftarrow l_w$; go to Step 2.

---

fixed, and then it recomputes l_b with l_w fixed. The iteration continues as long as there are improvements. How to recompute l_b with l_w fixed is straightforward. We show how to recompute l_w with l_b fixed while jointly optimizing the cost of both. The difficulty here is that *the cost of the backup path l_b changes as the working path l_w changes.*

The basic idea is to consider the changes in backup cost when recomputing the working path l_w. To recompute l_w, we use a standard shortest-path algorithm (such as Dijkstra's algorithm) with a modified *relaxation step* [Cormen et al., 1990] outlined below.

Let H_b be the number of hops l_b has. For any link $e \in E$, associate a backup cost vector $B_e = \{B_e^h | 1 \leq h \leq H_b\}$. B_e^h denotes the cost of l_b's h^{th} hop if e is used by l_w. B_e^h is defined as follows (e' denotes the hth hop of l_b in the definition below):

$$B_e^h := \begin{cases} +\infty & \text{if } \nu_{e'}^e = \nu_{e'}^* \wedge \lambda_f^e = 0 \\ C(e') & \text{if } \nu_{e'}^e = \nu_{e'}^* \wedge \lambda_f^e > 0 \\ \epsilon \cdot C(e') & \text{otherwise} \end{cases}$$

The first case ($\nu_{e'}^{e} = \nu_{e'}^{*} \wedge \lambda_f^e = 0$) indicates that link e' needs to allocate a free wavelength to protect a new working path traversing link e. As there is no free wavelength on link e', link e cannot be used by the working path l_w (l_b is fixed). The second case ($\nu_{e'}^{e} = \nu_{e'}^{*} \wedge \lambda_f^e > 0$) implies that link e' needs to allocate a free wavelength (and it is available) to protect a new working path traversing link e. The cost of link e' in this case is the original cost. The last case implies that link e' does not need to allocate any more free wavelength to protect one more working path traversing link e. Thus, link e' is used for "free" if link e is the working path. (Please refer to Fig. 2.2.)

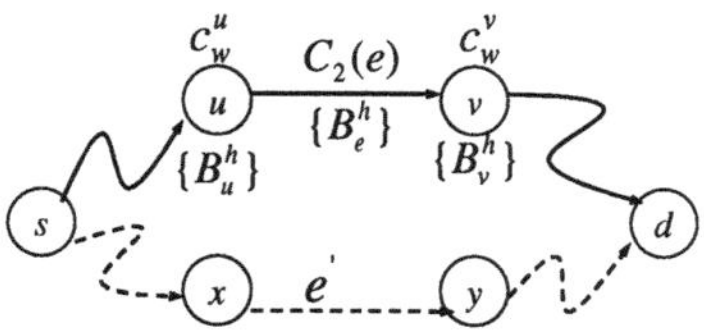

Figure 2.2. Illustration of OPT (dashed line is the fixed backup).

We redefine the link-cost function for computing the working path as follows:

$$C_2(e) := \begin{cases} +\infty & \text{if } e \in l_b \vee \lambda_f^e = 0 \\ C(e) & \text{otherwise} \end{cases}$$

If l_b traverses link e or there is no free wavelength on link e, then the working path cannot utilize link e and its cost is infinite; otherwise, the cost of link e is its original cost.

For any node $v \in V$, associate a cost variable c_w^v and a backup cost vector $B_v = \{B_v^h | 1 \le h \le H_b\}$. c_w^v denotes the cost of the working path from s to v. $\sum_h B_v^h$ indicates the backup cost if the minimum-cost path from s to v is used by the working path l_w. By definition, $C_w(l_w) = c_w^d$ and $C_b(l_w, l_b) = \sum_h B_d^h$.

We then employ a standard shortest-path algorithm to minimize $C_w(l_w) + C_b(l_w, l_b) = c_w^d + \sum_h B_d^h$ with the following relaxation step (please refer to Fig. 2.2):

RELAX(u, v)

LET $B^h = max\{B_u^h, B_e^h\}$, $1 \le h \le H_b$

IF $c_w^u + C_2(e) + \sum_h B^h < c_w^v + \sum_h B_v^h$ THEN

$c_w^v \leftarrow c_w^u + C_2(e)$

$B_v^h \leftarrow B^h$, $1 \le h \le H_b$

SET NODE u AS NODE v'S PREVIOUS HOP

The idea of the above relaxation step is to "relax" along the path which leads to the minimum total cost of the working and the backup paths. To decide whether to use node u as node v's previous hop along the working path (if node v is traversed by the working path), the relaxation step compares the total cost of the working ($c_w^u + C_w(w)$ vs. c_w^v) and the backup ($\sum_h B^h$ vs. $\sum_h B_v^h$). Thus, the changes in backup cost when the working path changes are correctly captured.

A formal specification of OPT is in Algorithm 2.2. Clearly, if l_w or l_b hits optimum in one iteration, OPT will stop at the next iteration and output the joint optimal l_w and l_b.

Algorithm 2.2 OPT

Input: $G = (V, E, C_2, \lambda)$, $\nu = \{\nu_e \mid e \in E\}, s, d \in V, l_w, l_b$
Output: Optimized l_w and l_b.

1 compute B_e for $e \in E$ with l_b fixed,

2 $c_w^s \leftarrow 0, c_w^v \leftarrow +\infty \ (v \in V \wedge v \neq s)$;
$B_s^h \leftarrow 0, B_v^h \leftarrow +\infty \ (1 \leq h \leq H_b, v \in V \wedge v \neq s)$,

3 compute a minimum-cost path as $l_w^{'}$ using a standard shortest-path algorithm with the new relaxation step shown above,

4 compute a minimum-cost path as $l_b^{'}$ with $l_w^{'}$ fixed using cost function C_1 in Algorithm 2.1, and

5 if $C_w(l_w^{'}) + C_b(l_w^{'}, l_b^{'}) < C_w(l_w) + C_b(l_w, l_b)$ then $l_w \leftarrow l_w^{'}, l_b \leftarrow l_b^{'}$, go to Step 1; otherwise return l_w and l_b.

The computational complexity of OPT is $O(|E|^2)$. In particular, the complexities of Steps 1-5 are $O(|E|^2)$, $O(1)$, $O(|V|^2)$, $O(|E|^2)$, and $O(1)$, respectively; and the number of iterations is typically a small constant for practical-sized networks.

OPT differs from Algorithm L in [Doshi et al., 1999] in the way the working path is computed for a given backup path. Since the cost of a backup path correlates to its working path, both OPT and Algorithm L in [Doshi et al., 1999] consider the changes in backup cost when computing the working path for a given backup path. However, in [Doshi et al., 1999], the changes in backup cost were considered to be *additive* for each hop along the working path. For example, if the first hop of a candidate working path incurs c_1 backup cost and the second hop of the candidate working path incurs c_2 backup cost, Algorithm L in [Doshi et al., 1999] considers the changes in backup cost to be $c_1 + c_2$ for the two hops along the candidate working path. However, the changes in backup cost is more meaningful if it is stated as $max\{c_1, c_2\}$, which is the case in OPT. The change from a linear relation, additive, to a non-linear relation, max, considerably affects how a shortest-path algorithm is modified to compute

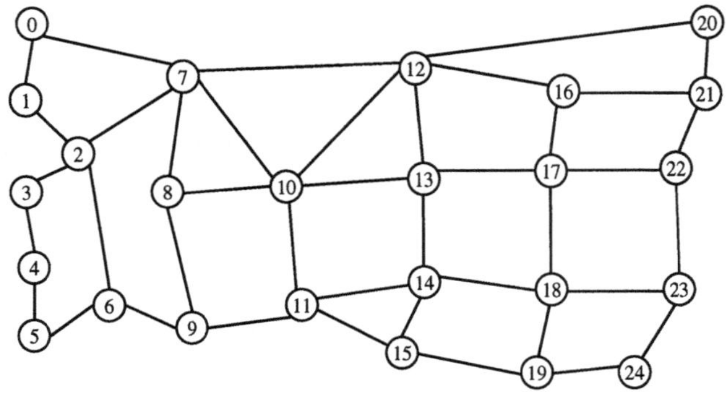

Figure 2.3. An example network used in this study.

the working path. In the case of Algorithm L in [Doshi et al., 1999], there is no need to modify a shortest-path algorithm (only to redefine link cost). In the case of OPT, a shortest-path algorithm needs to be modified and a vector of link cost needs to be redefined. Both OPT and Algorithm L in [Doshi et al., 1999] have complexity $O(|E|^2)$, but OPT has a larger constant factor.

2.5 Illustrative Numerical Results

We now quantitatively evaluate our heuristic algorithms. We simulate a dynamic network environment with the assumptions that the lightpath-arrival process is Poisson and the lightpath-holding time follows a negative exponential distribution. For the illustrative results shown here, in every experiment, 10^6 lightpath requests are simulated; they are uniformly distributed among all node pairs; average lightpath-holding time is normalized to unity; the cost of any link is unity; and our example network topology with 16 wavelengths per fiber is shown in Fig. 2.3.

We compare our heuristics to an effective two-step approach called Full Information Routing (FIR) [Li et al., 2002]. For a lightpath request, FIR first computes a least-cost path as the working path and then computes as backup path a least-cost path using cost function $C_1(e)$ (without the load balancing) based on the working path. The feasible solution to OPT is the solution of CAFES (with $k = 1$ since we found that the performance improvement is marginal if we increase k to any larger value)[4].

[4]While we use CAFES to generate feasible solutions as input for OPT here, we can use any shared-path-protected routing heuristic, e.g., FIR. In fact, we used FIR to generate feasible solutions as input to OPT, and we observed modest reduction in blocking probability. Thus, OPT is complementary to existing shared-path-protected routing approaches.

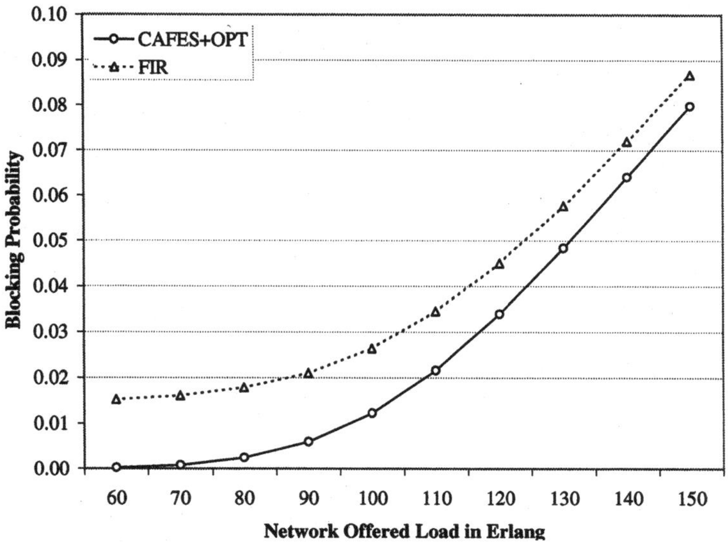

Figure 2.4. Blocking probability.

2.5.1 Blocking Probability

Figure 2.4 compares the blocking probability of our heuristics to FIR. We observe that our heuristic has lower blocking probability. This is because of the backtracking in CAFES and the joint optimization in OPT. The reason that the difference between the two curves decreases as the network offered load increases is as follows. When the network offered load is modest or low, a lightpath is unlikely to be blocked because of resource limitation. Blocking happens mainly because a heuristic cannot find two eligible link-disjoint paths. Since FIR fails in trap situations while CAFES does not, FIR has a much higher blocking probability. When the network offered load is high, a lightpath is more likely to be blocked because there are no available wavelengths on some links. While some lightpath requests still get blocked due to trap situations in FIR, other lightpath requests (which might be blocked in CAFES) can utilize the resources these lightpath requests are supposed to use (if they are accepted). Although CAFES can find a solution for a lightpath request with much higher probability than FIR, accepting a lightpath with a "detoured" route (as in the case when FIR cannot find a route while CAFES can) might interfere or even block future lightpath arrivals. Thus, the difference between the two curves decreases (but they will not cross each other because of the joint optimization in OPT).

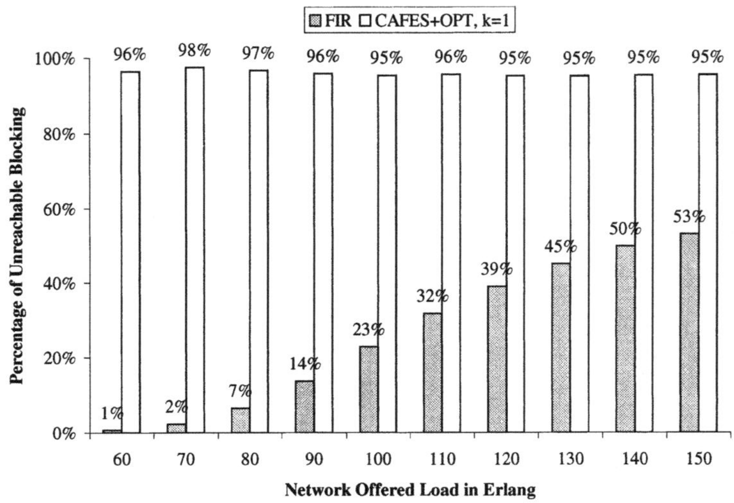

Figure 2.5. Percentage of unreachable blocking.

2.5.2 Percentage of Unreachable Blocking

The percentage of unreachable blocking (PUB) is defined as the percentage of blocking due to the fact that a pair of link-disjoint paths does not exist for a lightpath under the current network state. The other type of blocking is that, due to the NP-completeness of the problem, a heuristic may not find two link-disjoint paths even though they exist. Intuitively, this metric measures how "unlikely" it is that a lightpath request will be blocked when an eligible solution exists. Due to the NP-completeness of the problem, there is no polynomial-time algorithm to compute this metric. Here, we use an approximation. The approximate PUB is defined as the percentage of blocking we are sure that no link-disjoint paths exist for a lightpath under the current network state. (Clearly, the approximate PUB is slightly less than the exact PUB.) For example, in FIR, this metric is the percentage of blocking due to working paths not found. In our heuristics, for $k > 0$, it is the percentage of blocking that happens at Step 2 in Algorithm 2.1, i.e., the percentage of blocking due to working paths not found and due to the situation that the l_w in the $(k-1)^{th}$ iteration is the same as the l_w in the kth iteration. As shown in Fig. 2.5, with only one iteration ($k = 1$), more than 95% blocking is due to unreachability, compared to less than 60% in FIR for the heavy-load region (150 Erlangs). The main reason why FIR has such a low PUB is that there exists a trap situation in the topology: FIR cannot find two link-disjoint paths from node 9 to node 3 because: 1) FIR computes as working path the least-cost path $\langle 9, 6, 2, 3 \rangle$; 2) and there is no path link-disjoint to the working path $\langle 9, 6, 2, 3 \rangle$ from node 9 to node 3. Another reason is that a non-trap topology can later become a trap topology after all the wavelengths

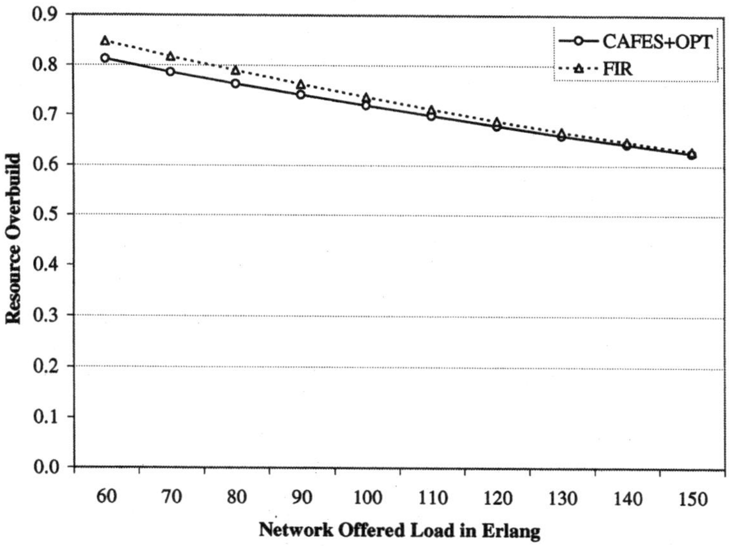

Figure 2.6. Resource overbuild.

on some link are used up. Further results (not shown here) from a non-trap topology show that our heuristics have much higher PUB compared to FIR, though the difference is not as drastic as here. If we increase k, then the PUB in our heuristics will further approach towards 100%.

2.5.3 Resource Overbuild

One figure of merit for comparing resource efficiency is *resource overbuild*, defined as the amount of wavelength links consumed by backup paths over the amount of wavelength links utilized by working paths [Li et al., 2002]. Resource overbuild indicates the amount of extra resources needed for providing protection as the percentage of the amount of resources required without protection. Typically, it is desirable to have lower resource overbuild because lower resource overbuild implies better backup sharing. Figure 2.6 shows that our heuristics have lower resource overbuild over FIR. The optimization procedure in OPT contributes to the reduction in resource overbuild.

2.5.4 Average Hop Distance

The advantages of our heuristics come for a slight increase in the average backup-path hop distance. Figure 2.7 shows that both our heuristics and FIR have similar average working-path hop distance while our heuristics have slightly longer average backup-path hop distance. The load balancing in CAFES and the joint optimization in OPT (which tends to maximize backup shareabil-

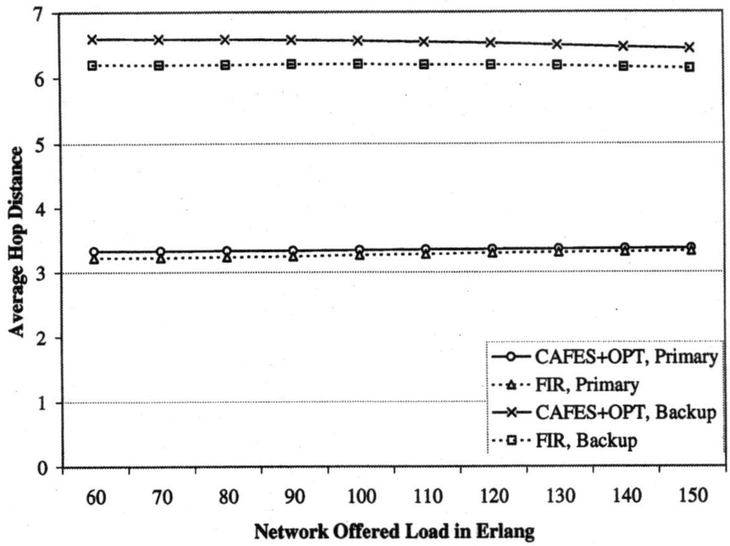

Figure 2.7. Average hop distance.

ity) lead to the increase in backup hop distance. Figures 2.4, 2.6, and 2.7 imply that our heuristics lead to better backup sharing than FIR does because: 1) the average backup-path hop distance of our heuristics is longer; 2) our heuristics have lower resource overbuild; 3) and our heuristics have lower blocking probability.

2.6 Conclusion

This chapter is devoted to dynamic shared-path-protected lightpath provisioning in optical WDM mesh networks. We proved that the problem of finding an eligible pair of working and backup paths under shared-path-protection constraints in a WDM mesh network for a new lightpath request with respect to existing lightpaths is NP-complete. We presented a heuristic, called CAFES, to compute a feasible solution and another heuristic, called OPT, to optimize the resource consumption for a given solution. Simulation results demonstrated that our heuristic algorithms are efficient and effective. While our focus here is on dynamic lightpath provisioning in WDM context, the core ideas of how to compute a feasible solution and how to jointly optimize a given solution can be applied to: 1) restorable bandwidth-guaranteed-connection provisioning in MPLS networks; and 2) static shared-path-protected lightpath provisioning in which the traffic matrix is known a priori.

In the next chapter, we shall investigate not only the resource efficiency of shared-path protection, but also the fault-recovery time and scalability.

APPENDIX 2.A: NP-Completeness of DSPLP Problem

Basic idea on our proof of Theorem 2.1: We reduce 3SAT, which is known to be NP-complete [Garey and Johnson, 1979], to the Dynamic Shared-Path-Protected Lightpath-Provisioning (DSPLP) problem. The 3SAT problem is formally stated as follows.

3SAT Instance: Collection $F = \{D_1, D_2, \ldots, D_M\}$ of clauses on a finite set Q of variables such that $|D_i| = 3$ for $1 \leq i \leq M$, where clause D_i is the boolean "or" of three literals (a literal is either a variable or the boolean "not" of a variable) and is satisfied by a truth assignment if and only if at least one of the three literals is true. $|F| = M$, where M is the number of clauses; $|Q| = N$, where N is the number of variables.

3SAT Question: Is there a truth assignment for Q that satisfies all the clauses in F?

The basic idea of our proof is, for an arbitrary instance of 3SAT, to construct a graph such that there is a path l_w corresponding to the truth assignment of all the clauses and a path l_b corresponding to the false assignment of all the variables. The two paths satisfy the shared-path-protection constraints (Constraints C.1-C.4) if and only if the given instance of 3SAT is satisfied. The formal proof follows.

Proof of Theorem 2.1: DSPLP $\in$ NP since a non-deterministic algorithm can guess two paths, l_w and l_b, and check in polynomial time if these two paths satisfy Constraints C.1-C.4 with respect to existing lightpaths in $\mathcal{L}$.

Given an arbitrary instance of 3SAT $F = \{D_1, D_2, \ldots, D_M\}$ and $Q = \{v_1, v_2, \ldots, v_N\}$, we construct in polynomial time an instance of DSPLP $G = (V, E, C, \lambda)$, $\mathcal{L}$, s, and d where $C(e) = 1$ and $\lambda(e) = 1$ for all $e \in E$. In the following construction, we first define the set of nodes V, then define the set of links E, and finally define the set of existing lightpaths $\mathcal{L}$.

The set of nodes V consists of a source node s and a destination node d. The other nodes in V can be divided into three groups. The first group is related to the variables in Q and consists of the nodes $a_k, b_j^i, \overline{b}_j^i, c_j^i$, and $\overline{c}_j^i$, where $0 \leq k \leq N, 1 \leq i \leq M, 1 \leq j \leq N$. The second group of nodes is related to the clauses in F and consists of the nodes p_i and q_i for $1 \leq i \leq M$. The third group of nodes is the source nodes of the existing lightpaths and consists of z_i for $0 \leq i \leq N$.

The set of links (unidirectional fibers) E can also be divided into three groups. In the reduction, the first group of links will be used for assigning values to the variables in Q; the second group of links will be used for evaluating the boolean value of the clauses; the third group of links will be used for existing lightpaths. The three groups of links are as follows.

- Group 1:
  - (a) A link from node b_j^i to node c_j^i, $1 \leq i \leq M$ and $1 \leq j \leq N$.
  - (b) A link from node $\overline{b}_j^i$ to node $\overline{c}_j^i$, $1 \leq i \leq M$ and $1 \leq j \leq N$.
  - (c) A link from node c_j^i to node b_j^{i+1}, $1 \leq i \leq M-1$ and $1 \leq j \leq N$.
  - (d) A link from node $\overline{c}_j^i$ to node $\overline{b}_j^{i+1}$, $1 \leq i \leq M-1$ and $1 \leq j \leq N$.
  - (e) A link from node a_j to node b_{j+1}^1, and a link from node a_j to node $\overline{b}_{j+1}^1$, $0 \leq j \leq N-1$.
  - (f) A link from node c_j^M to node a_j, and a link from node $\overline{c}_j^M$ to node a_j, $1 \leq j \leq N$.
- Group 2:
  - (a) A link from node p_i to b_j^i, and a link from node c_j^i to node q_i, if and only if the variable v_j is in clause D_i.
  - (b) A link from node p_i to $\overline{b}_j^i$, and a link from node $\overline{c}_j^i$ to node q_i, if and only if $\overline{v}_j$ is in clause D_i.

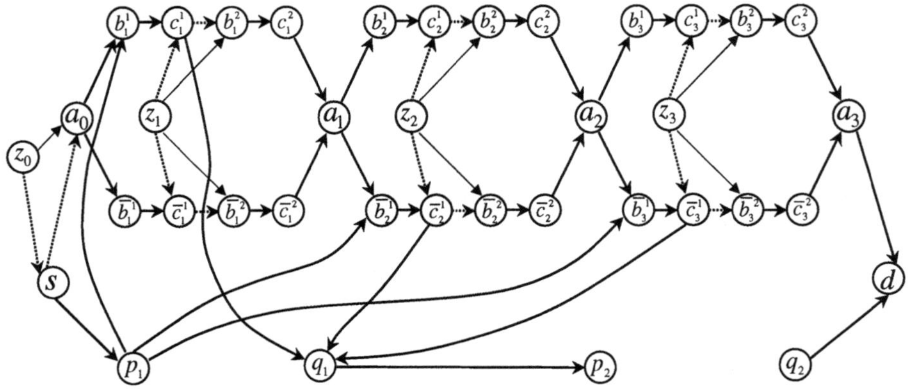

Figure 2.A.1. An illustrative construction for a 3SAT instance $F = \{D_1, D_2\}, D_1 = v_1 \vee \bar{v}_2 \vee \bar{v}_3, D_2 = \bar{v}_1 \vee v_2 \vee v_3$, and $Q = \{v_1, v_2, v_3\}$ (the part corresponding to clause D_2 is not shown). A dotted line corresponds to a link whose only wavelength has been used by an existing backup path. A thin solid line corresponds to a link whose only wavelength has been used by an existing working path. A thick solid line corresponds to a link whose only wavelength is available.

(c) A link from node q_i to node p_{i+1}, $1 \leq i \leq M-1$.

- Group 3:

(a) A link from z_0 to s and a link from z_0 to a_0.

(b) A link from z_j to c_j^i and a link from z_j to $\bar{c}_j^i$, where $1 \leq i < M$ and $1 \leq j \leq N$.

(c) A link from z_j to b_j^i and a link from z_j to $\bar{b}_j^i$, where $1 < i \leq M$ and $1 \leq j \leq N$.

In addition to the above links, we also have a link from node s to node a_0, a link from node a_N to node d, a link from node s to node p_1, and a link from node q_M to node d. It is easy to see that the construction can be done in polynomial time.

The set of $2 \times (M-1) \times N + 1$ existing lightpaths $\mathcal{L}$ is defined as follows:

1 Lightpath 1: working path $\langle z_0, a_0 \rangle$, backup path $\langle z_0, s, a_0 \rangle$.

2 Lightpaths 2 to $(M-1) \times N + 1$: working path $\langle z_j, b_j^{i+1} \rangle$, backup path $\langle z_j, c_j^i, b_j^{i+1} \rangle$, where $1 \leq i < M$ and $1 \leq j \leq N$.

3 Lightpaths $(M-1) \times N + 2$ to $2 \times (M-1) \times N + 1$: working path $\langle z_j, \bar{b}_j^{i+1} \rangle$, backup path $\langle z_j, \bar{c}_j^i, \bar{b}_j^{i+1} \rangle$, where $1 \leq i < M$ and $1 \leq j \leq N$.

An illustrative construction (the construction for the second clause is similar to the first one and thus it is not shown) for a 3SAT instance $F = \{D_1, D_2\}, D_1 = v_1 \vee \bar{v}_2 \vee \bar{v}_3, D_2 = \bar{v}_1 \vee v_2 \vee v_3$, and $Q = \{v_1, v_2, v_3\}$ is shown in Fig. 2.A.1.

We now show that, if F is satisfiable, then from node s to node d in graph G there exist two paths satisfying the shared-path-protection constraints C.1-C.4 with respect to the existing lightpaths $\mathcal{L}$. Let $v_1 = x_1, v_2 = x_2, \ldots, v_N = x_N$ be an assignment that satisfies F, where $x_j \in \{0, 1\}$ for $1 \leq j \leq N$. The two paths can be routed as follows. The backup path l_b is routed via the nodes b_j^i and c_j^i $(1 \leq j \leq N$ and $1 \leq i \leq M)$ if and only if $x_j = 0$; otherwise, l_b is routed via nodes $\bar{b}_j^i$ and $\bar{c}_j^i$ (l_b will also need to traverse all the a_k nodes for $0 \leq k \leq N$). The working path l_w is routed via p_i, q_i, and other nodes defined as follows. By the construction, link $\langle p_i, q_i \rangle$ corresponds to clause $D_i, |D_i| = 3$. Without loss of generality, let $D_i = v_f \vee \bar{v_g} \vee v_h$, where f, g, and h are distinct integers between 1 and N. Then, there will be three paths from node p_i to node q_i that go through the nodes $b^i_f, \bar{b}^i_g$, and b^i_h, respectively. Since $v_1 = x_1, v_2 = x_2, \ldots, v_N = x_N$ is an assignment that satisfies F, either $x_f = 1$, or $x_g = 0$, or $x_h = 1$. If $x_f = 1$, then the working path traverses nodes b^i_f and c^i_f; if $x_g = 0$, then the working path traverses nodes $\bar{b}^i_g$ and $\bar{c}^i_g$; if $x_h = 1$, then the working path traverses nodes b^i_h and c^i_h; if more than one condition is true, then randomly pick one. l_w and l_b so selected are link disjoint (Condition C.1) because of the following facts:

1 l_w traverses links in Group 1 and Group 2.

2 l_b only traverses links in Group 1.

3 The Group-1 links l_w traverses correspond to the literals of value 1. We will use the term "1-value link" ("0-value link") to denote a link corresponding to a literal of value 1 (0).

4 The Group-1 links l_b traverses correspond to the literals of value 0.

5 A Group-1 1-value link is disjoint to a Group-1 0-value link by the construction.

6 A Group-1 link is disjoint to a Group-2 link by the construction.

The Condition C.2 holds because: l_w uses links in Group 1 and Group 2 while any working path in $\mathcal{L}$ only utilizes links in Group 3; a link in Group 3 is disjoint to a link in either Group 1 or Group 2. The Condition C.3 holds because: any backup path in $\mathcal{L}$ utilizes 0-value links in Group 1 and links in Group 3; l_w uses 1-value links in Group 1 and links in Group 2; a link in Group 2 is disjoint to a link in either Group 1 or Group 3; a 0-value link in Group 1 is disjoint to a 1-value link in Group 1. The Condition C.4 holds because l_w does not traverse the same SRLG as any working path in $\mathcal{L}$ (thus, l_b can share any wavelength-link with any backup path in $\mathcal{L}$) due to the facts that: l_w only utilizes links in Group 1 and Group 2; any working path in $\mathcal{L}$ only utilizes links in Group 3; a link in Group 3 is disjoint to a link in either Group 1 or Group 2. Thus, we have shown that, if F is satisfiable, then we can find for the lightpath request from s to d two lightpaths satisfying shared-path-protection constraints with respect to the existing lightpaths.

Suppose from s to d in the constructed graph G there exist two paths (working path l_w and backup path l_b) satisfying the shared-path-protection constraints (Constraints C.1-C.4) with respect to the existing lightpaths $\mathcal{L}$. We show that F is satisfiable using the following facts:

1 Since there are only two links emanating from the source node s, one of them will be used by the working path l_w while the other will be used by the backup path l_b due to Constraint C.1 that l_w and l_b are link-disjoint.

2 The working path l_w is routed via the p_i and q_i nodes (and some Group-1 nodes which will be specified later) because of Constraint C.2 and the construction that the only wavelength on link $\langle s, a_0 \rangle$ is used by an existing backup path.

3 The backup path l_b traverses links only in Group 1 and all the a_j nodes. If the backup path l_b uses any link in Group 2, then l_b will have to traverse some link $\langle q_i, p_{i+1} \rangle$ for some $1 \leq i < M$, which is used by the working path. Thus, the two lightpaths will violate Constraint C.1. Clearly l_b cannot use any link in Group 3. Furthermore, if, for some value of j, the backup path traverses node b^i_j ($\bar{b}^i_j$), then it must traverse all the b^i_j ($\bar{b}^i_j$) nodes and the c^i_j ($\bar{c}^i_j$) nodes, where $1 \leq i \leq M$.

4 We represent the route for the backup path l_b by a N-bit binary number, $\beta_1\beta_2\ldots\beta_N$. The backup path is routed as follows: if bit β_j is 1, then the backup path is routed over the nodes $\bar{b}^i_j$; otherwise, it is routed over the nodes b^i_j. Note that there is a one-to-one mapping between an N-bit binary number and a backup route in the network. Let $\beta_1\beta_2\ldots\beta_N$ be the N-bit number that corresponds to the backup path. Then, there is a path from node p_i to

node q_i that does not use the links in the backup path, if and only if clause D_i is true under the following assignment: $v_j = \beta_j$, where $1 \leq j \leq N$. The reason is as follows. Without loss of generality, let $D_i = v_f \vee \overline{v}_g \vee v_h$, where f, g, and h are distinct integers between 1 and N. Note that there are three paths from node p_i to node q_i that traverse the nodes b_f^i, $\overline{b}_g^i$, and b_h^i, respectively. The backup path will traverse the nodes b_f^i, $\overline{b}_g^i$, and b_h^i, if and only if $\beta_f = 0$, $\beta_g = 1$, and $\beta_h = 0$, respectively. Thus, there exists a path from node p_i to q_i if and only if β_f is 1, or β_g is 0, or β_h is 1, which is exactly the condition under which D_i will be true.

Since the working path and the backup path satisfy the shared-path-protection constraints, D_i is true for all $1 \leq i \leq M$ due to the above facts. Thus, F is satisfiable under the following assignment: $v_j = \beta_j, 1 \leq j \leq N$, where $\beta_1\beta_2\ldots\beta_N$ is the N-bit number corresponding to the backup path l_b.

This concludes our proof that DSPLP is NP-complete.

Chapter 3

SUB-PATH PROTECTION FOR SCALABILITY AND FAST RECOVERY

3.1 Introduction

As networks migrate from stacked rings to meshes because of the poor scalability of interconnected rings and the excessive resource redundancy used in ring-based protection [MacDonald et al., 2000], mesh-structured protection schemes have been receiving increasing attention [Doshi et al., 1999, Ramamurthy et al., 2003, Iraschko et al., 1998, Van Caenegem et al., 1998, Crochat et al., 2000, Miyao and Saito, 1998, Mohan et al., 2001]. We review the work on WDM mesh protection for a given set of lightpath requests (which is the focus of this chapter), and classify them based on whether they treat the underlying mesh as a whole, or they fragment the mesh into other protection domains [Gerstel and Ramaswami, 2000a], or they split an end-to-end lightpath into different segments.

3.1.1 Related Work

The first category of work, as in [Doshi et al., 1999, Ramamurthy et al., 2003, Iraschko et al., 1998, Van Caenegem et al., 1998, Crochat et al., 2000, Miyao and Saito, 1998], proposes different protection schemes to protect the underlying mesh network as a whole. Specifically, the work in [Ramamurthy et al., 2003] develops integer linear programs (ILPs) for RWA with dedicated-path protection, shared-path protection, and shared-link protection against single-link failures. The objective is to minimize the total number of wavelength-links, where a wavelength-link is a wavelength on some link. The work in [Doshi et al., 1999] considers two problems: determining the best backup route for each lightpath request, given the network topology, the capacities, and the primary routes of all requests; and determining primary and backup routes for each lightpath request to minimize network capacity and cost. ILP and distributed heuristic algorithms, under single-link or single-node failures, are presented. The work in [Iraschko et al., 1998] jointly optimizes primary and backup paths for path protection against single-link failures with failure-dependent protection. Lower bounds on spare-capacity requirements and integer-program formulations are presented. The work in [Van Caenegem et al., 1998] investigates the problem of routing, planning of primary capacity, rerouting, and planning of spare capacity in WDM networks. An ILP and a simulated-annealing-based heuristic are used to minimize the total cost for a given static traffic demand. The work in [Miyao and Saito, 1998] formulates the RWA problem under dedicated-path and shared-path-protection constraints into integer programs, whose objective is to minimize the total facility cost, including both transmission and crossconnect cost. The work in [Crochat et al., 2000] considers protection interoperability between optical layer (e.g., WDM transmission layer) and higher layer (e.g., IP layer). The authors design a network mapping between the two layers such that it respects the capacity constraints, survives from single-link or single-node failures, and ensures the number of lightpaths affected by such a failure is less than some given constant.

The second category of work, presented in [Ellinas et al., 2001, Fumagalli et al., 1999, Gardner et al., 1994, Grover and Stamatelakis, 1998, Medard et al., 2002, Medard et al., 1999], decomposes a mesh network into different structures, such as rings, protection cycles, digraphs, pre-configured cycles (p-cycles), or trees. Specifically, the work in [Ellinas et al., 2001, Gardner et al., 1994] decomposes a mesh into 4-fiber rings (which [Ellinas et al., 2001] refers to as protection cycles), which then perform automatic protection switching (APS). The work in [Medard et al., 2002] creates primary and secondary digraphs based on a mesh so that the secondary digraph can be used to carry backup traffic that provides loop-back to the primary graph upon failures. The work in [Grover and Stamatelakis, 1998] proposes the use of pre-configured cycles, or p-cycles, where a cycle protects not only the lightpaths that are part of it, but also chords that run between cycle nodes. The work in [Medard et al., 1999] creates redundant trees on arbitrary node-redundant or link-redundant networks to combat against single-link or single-node failures. The work in [Fumagalli et al., 1999] presents ILPs to decompose a WDM mesh network into self-healing rings.

The third category of work in [Anand et al., 2002, Gummadi et al., 2003, Ho and Mouftah, 2002] addresses mesh-structured protection against single-link failures by dividing a primary path into a sequence of segments and protecting each such segment separately. In particular, the approach in [Anand et al., 2002] divides a primary path into disjoint segments; the work in [Gummadi et al., 2003, Ho and Mouftah, 2002] divides primary paths into overlapped segments, thus the network survives from single-node failures.

By treating the underlying mesh network as a whole, the work in the first category can achieve optimal resource utilization since it has complete information on the entire network. It may, however, lead to long protection-switching time, and scalability can become a significant issue as the size of network increases. The work in the second category decomposes a mesh network into different types of protection structures and then applies APS or self-healing ring (SHR). While this may be a short-term solution for accommodating legacy ring algorithms and equipment, it may lead to excessive resource redundancy. As pointed out in [MacDonald et al., 2000], the minimum redundancy for a 1:1/DP (Diverse Protection) APS system is 100% and the redundancy of a SHR network is at least 100% and up to 300%. Because backup paths cannot be computed without knowledge of the segmentation points along the primary paths, the work in [Anand et al., 2002, Ho and Mouftah, 2002] computes primary paths prior to backup paths, which has been shown to be resource inefficient in [Iraschko et al., 1998, Bhandari, 1999, Sen et al., 2001, Xin et al., 2001a]. While [Gummadi et al., 2003] dynamically divides a primary path into multiple segments, it does not accommodate backup sharing.

3.1.2 Multi-Domain Optical Networks and Our Proposal

Optical networks are expected to operate as multiple routing domains due to technological constraints, administrative functions, trust relationship, and other considerations [Strand and Xue, 2001]. An optical network domain is defined in [Strand and Xue, 2001] as "a portion of an interconnected optical network that has a clear demarcation boundary based on technology, business, service, technical administration, or architectural function." A group of products from the same vendor may consist of an optical network domain. Metro/core optical networks may comprise a multi-domain optical network [Wang et al., 2002a]. As optical networks evolve to multi-vendor heterogeneous networks, multiple optical network domains are likely to be needed [Strand and Xue, 2001, Chiu and Strand, 2003].

We propose a new protection scheme, called sub-path protection, which combines the optical-domain idea with shared-path protection to achieve high scalability and fast recovery time for a modest sacrifice in resource utilization. *The main ideas of sub-path protection are (1) to partition a large optical network into several smaller domains, and then (2) to apply shared-path protection to the optical network such that an intra-domain lightpath does not use resources of other domains and the primary/backup paths of an inter-domain lightpath exit a domain (and enter another domain) through a common egress (or ingress) domain-border node (DBN).*

While sub-path protection assumes the concept of shared-path protection underneath for routing within each domain, it can utilize other protection schemes such as dedicated-path protection as well. Sub-path protection can even employ different protection schemes in different domains to provide differentiated quality of service (QoS) based protection for lightpath requests with differentiated availability requirements [Wen et al., 2002].

In this chapter, we assume that the optical network under investigation belongs to one single carrier since inter-carrier-network communication is typically achieved via policy-based protocols such as border gateway protocol (BGP) on IP layer. We further concentrate on single-fiber[1] failures since they are the predominant form of failures in optical networks[2].

3.1.3 Organization

The rest of this chapter is organized as follows. Section 3.2 discusses the different variants of sub-path protection, proposes an approach to handle single-node failures, formally defines the problem of routing and wavelength-assignment (RWA) under sub-path protection, and proves that it is NP-complete. Section 3.3 mathematically formulates the RWA problem under sub-path protection for a given set of lightpath requests. Section 3.4 proves that the problem of finding optimal backup paths, under sub-path-protection constraints (or shared-path-protection constraints) for a set of lightpath requests whose primary paths are given, is still NP-complete. This section also develops a heuristic to find efficient solutions. Section 3.5 compares the fault-recovery time, survivability, scalability, and resource utilization of sub-path protection with those of shared-path protection. Section 3.6 concludes this chapter.

3.2 Sub-Path Protection

3.2.1 An Illustrative Example

Consider an inter-domain lightpath between node-pair $(3, 16)$, as shown in Fig. 3.1. The primary path, p_p, is $\langle 3, 4, 7, 14, 16 \rangle$, and the backup path, p_b, is $\langle 3, 6, 5, 7, 11, 14, 17, 16 \rangle$. Note that p_p and p_b exit Domain 1 (and enter Domain 2) from the same DBN 7 and exit Domain 2 (and enter Domain 3) from the same DBN 14. DBNs 7 and 14 segment p_p and p_b into three sub-paths, respectively, $\langle 3, 4, 7 \rangle$ (p_{p_1}), $\langle 7, 14 \rangle$ (p_{p_2}), and $\langle 14, 16 \rangle$ (p_{p_3}) for p_p; $\langle 3, 6, 5, 7 \rangle$ (p_{b_1}), $\langle 7, 11, 14 \rangle$ (p_{b_2}), and $\langle 14, 17, 16 \rangle$ (p_{b_3}) for p_b. Each (p_{p_i}, p_{b_i}) pair corresponds to the primary and backup paths of some intra-domain segment of the lightpath. When link $\langle i, j \rangle$[3] (e.g., $\langle 7, 14 \rangle$) along p_p fails, instead of shutting

[1] In this chapter, a fiber is considered to be bidirectional, and the term "link" refers to a unidirectional fiber.
[2] Other failure scenarios may include shared-risk-link-group (SRLG) failures [Zang et al., 2003, Strand et al., 2001], and the more general shared-risk-group (SRG) failures. SRLG is an abstraction referring to a group of fibers that may be prone to a common failure. SRG is an abstraction referring to a group of network elements (including fibers, OXCs, and other network components) that may be subject to a common failure.
[3] Throughout this chapter, we use $\langle i, j \rangle$ and ij interchangeably to denote link $i \rightarrow j$: We use $\langle i, j \rangle$ in the text and ij in the formulations in later sections.

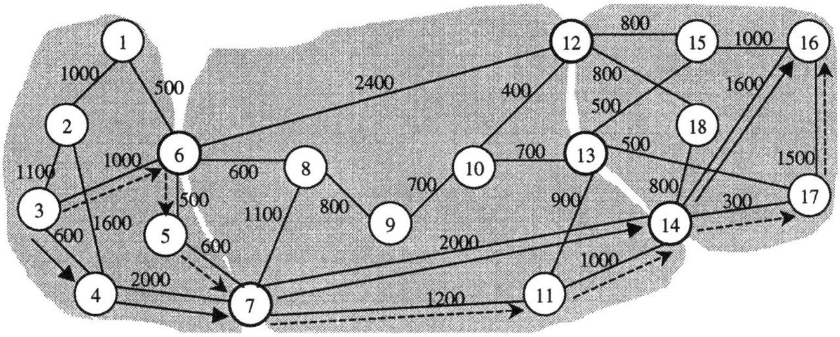

Figure 3.1. An example nationwide network where each cloud denotes a domain. Domain 1 includes nodes 1–7 and the links in between (if there is a link between two DBNs of two domains, the link belongs to one domain only); Domain 2 includes nodes 6–14 and the links in between; Domain 3 includes nodes 12–18 and the links in between. Nodes 6, 7, 12, 13, and 14 are DBNs. The number besides each link is the length of the link in kilometer. The solid (dashed) arrows form the primary (backup) path between node pair (3, 16).

down the entire primary path p_p and switching the traffic to p_b, our proposed scheme only turns down the sub-path p_{p_k} (p_{p_2} in this example) that traverses the failed link, and switches the traffic to p_{b_k} (p_{b_2}). As a result, other sub-paths (p_{p_1} and p_{p_3} in this example) that do not traverse link $\langle i, j \rangle$ are not affected.

3.2.2 Different Cases

Depending on the wavelength-conversion capability of DBNs, there are two cases of sub-path protection. In the absence of wavelength converters at DBNs, the primary and backup paths of an inter-domain lightpath must be on the same wavelength. Suppose DBNs 7 and 14 are wavelength continuous. If p_{p_1} is on wavelength w_1 and p_{b_2} is on w_2, when link $\langle 7, 11 \rangle$ fails, we cannot simply switch the traffic from p_{p_2} to p_{b_2} because p_{p_1} and p_{b_2} are on different wavelengths. If there are wavelength converters capable of full-range wavelength conversion at DBNs, then each sub-path of one lightpath can be on any wavelength, regardless of the wavelength assignments of other sub-paths of the same lightpath.

Depending on the administration of each domain, the implementation of DBNs may be different. Suppose that, in Fig. 3.1, Domain 1 and Domain 2 belong to different administrative entities. Then, in practice, DBN 6 (DBN 7 as well) is usually a logical node consisting of two nodes[4], where each node belongs to one administrative entity and all the physical nodes are inter-connected through high-capacity links. If both Domain 1 and Domain 2 belong to the same administrative entity, then in practice DBN 6 (DBN 7 as well) may be imple-

[4]Each of these nodes can again be a logical node consisting of multiple physical nodes for fault-tolerant purpose, as will be shown in Section 3.2.3.

mented using one node but it can be implemented using multiple nodes too. We assume that the links between the nodes of a DBN are $1 + 1$ protected [Wang et al., 2002a], as is usually done in practice, and treat all such nodes as one logical node in this study.

3.2.3 Domain-Border-Node (DBN) Failures

It may appear that sub-path protection may be susceptible to single-node failures at a DBN since the primary and backup paths of an inter-domain lightpath traverse common DBNs. Below we show one possible approach to dealing with single-node failures.

Consider an inter-domain lightpath between node pair (s, d). Suppose that the lightpath traverses DBN r. We can construct the DBN r by using two physical nodes r_1 and r_2 interconnected by a fiber, as shown in Fig. 3.2(a). (Figure 3.2(b) shows DBN 14 in Fig. 3.1 so constructed.) Below we show that the lightpath survives from any single fiber/node failures and still enjoys the benefits of sub-path protection, assuming that the two sub-paths between node s and DBN r (DBN r and node d) are fiber/node disjoint.

By the construction of DBN r, if link $\langle r_1, r_2 \rangle$ fails, the lightpath can take either path $\langle s, r_1, d \rangle$ or path $\langle s, r_2, d \rangle$ to carry traffic. Thus, the inter-domain lightpath survives from single-fiber failures.

By the construction of DBN r, if physical node r_1 fails, the lightpath can take path $\langle s, r_2, d \rangle$ to carry traffic; if physical node r_2 fails, the lightpath can take path $\langle s, r_1, d \rangle$ to carry traffic. Thus, the inter-domain lightpath survives from single-node failures.

Meanwhile, this construction allows an inter-domain lightpath to enjoy the benefits of sub-path protection, i.e., any single fiber/node failure only affects a segment of the entire lightpath. For example, if any fiber/node between node s and node r_1 fails, node s can switch the traffic to the path $\langle s, r_2 \rangle$; if node r_1 fails, node s can switch to the sub-path $\langle s, r_2 \rangle$; if node r_2 fails, node r_1 can switch to the sub-path $\langle r_1, d \rangle$; if any fiber/node between node r_2 and node d fails, r_1 can switch the traffic to the sub-path $\langle r_1, d \rangle$. In general, any single fiber/node failure only affects one domain.

While we may consider that nodes r_1 and r_2 employ 1+1 redundancy, we can deliberately group two nodes as a logical DBN to avoid redundancy; and sub-path protection can still function properly. For example, we can group nodes 6 and 7 as a logical DBN, and nodes 12 and 14 as another logical DBN in Fig. 3.1. In that case, the lightpath request between node pair $(3, 16)$ can have $\langle 3, 6, 5, 7, 14, 18, 12, 15, 16 \rangle$ as primary path. Its backup sub-paths can be $\langle 3, 4, 7 \rangle$, $\langle 6, 12 \rangle$, and $\langle 14, 17, 16 \rangle$. (There are details such as coordination between nodes in a logical DBN, but we are skipping them because they are relatively straightforward.)

Other possible ways of dual-node-interconnection approaches, such as Drop and Continue for SONET ring interconnection, can be found in [Foisel, 2001, Grover, 1999].

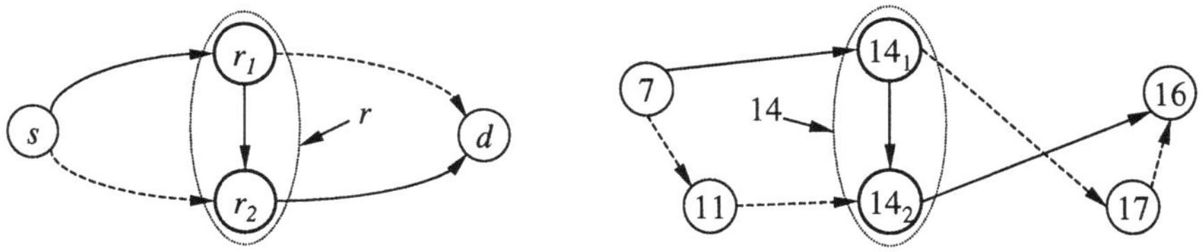

(a) Implementation of DBN r (b) Implementation of DBN 14 in Fig. 3.1

Figure 3.2. A possible DBN implementation to combat single-node failures. Primary (backup) paths are in solid (dashed) arrows.

3.2.4 Problem Statement

We first define the notations and then formally state the RWA problem with sub-path protection. A network is represented as a weighted, directed graph $G = (V, E, C)$, where V is the set of network nodes, E is the set of unidirectional links, and $C : E \rightarrow Z^+$ is the cost function for each link (Z^+ denotes the set of positive integers). A *domain* A on G is the node-induced subgraph $G_A = (V_A, E_A, C_A)$, where $V_A \subseteq V$, E_A contains all the links of E, both end points of which are in E_A, and $C_A : E_A \rightarrow Z^+$ is C restricting to E_A. A *partitioning configuration* $\mathcal{P}$ is a set of domains. A *domain-border node (DBN)* of $\mathcal{P}$ is a node that belongs to multiple domains of $\mathcal{P}$. The set of DBNs $\mathcal{R}$ for a partitioning configuration $\mathcal{P}$ is defined as $\mathcal{R} = \bigcup_{A,A' \in \mathcal{P}} (V_A \cap V_{A'})$. A partitioning configuration $\mathcal{P}$ is *proper* if each DBN of $\mathcal{P}$ has an in-degree of at least two in each domain it belongs to, and there is no link that crosses multiple domains. A lightpath-request set $\mathcal{D}$, $\mathcal{D} \subset V \times V$, is a set of source-destination pairs, where each pair represents a lightpath request from the source node to the destination node[5].

The RWA for the primary and backup paths of $\mathcal{D}$ satisfies the *shared-path-protection constraints* as follows:

C.1 The primary and backup paths of a lightpath are fiber-disjoint.

C.2 Two primary paths do not utilize the same wavelength on any common link they traverse.

[5] If there are multiple lightpath requests between the same source-destination pair, then $\mathcal{D}$ contains multiple copies of that source-destination pair.

C.3 A primary path does not share any wavelength with any backup path on any common link they traverse.

C.4 Two backup paths can share a wavelength on a common link only if their corresponding primary paths are fiber disjoint.

The RWA for the primary and backup paths of $\mathcal{D}$ satisfies the *domain constraints* as follows:

C.5 The primary and backup paths of an intra-domain lightpath k ($\exists\ A \in \mathcal{P}$ s.t. $k_s, k_d \in A$, where k_s and k_d are the source and destination nodes of lightpath k) only use the resources of its own domain (domain A).

C.6 The primary and backup paths of an inter-domain lightpath k ($\forall\ A \in \mathcal{P}, \wedge\neg(k_s, k_d \in A)$) exit a domain (and enter another domain) through a common DBN.

The RWA for the primary and backup paths of $\mathcal{D}$ satisfies the *sub-path-protection constraints* if it satisfies both shared-path-protection constraints and domain constraints.

We now formally state the RWA problem in an optical WDM mesh network with sub-path protection as follows. Given a network G with W wavelengths per link, a partitioning configuration $\mathcal{P}$, and a set of lightpath requests $\mathcal{D}$, route each lightpath on G and assign a wavelength to each path under sub-path-protection constraints such that the total network cost is minimized (which effectively maximizes the network throughput when network resources, e.g., number of wavelengths per link, are limited).

In this chapter, we assume the partitioning configuration of a network is given and it is proper. How to partition a large optical network into smaller domains involves administrative issues and is not addressed here. A related approach on partitioning a network can be found in [Hac, 1994]. We also assume that DBNs are wavelength convertible since, in practice, optical domains are expected to be isolated by transponders [Strand and Xue, 2001], making wavelength conversion available at DBNs.

3.2.5 Proof of NP-Completeness

While the static lightpath establishment (SLE) problem has been shown to be NP-complete for a wavelength-continuous network with $W \geq 3$ wavelengths [Chlamtac et al., 1992], so far there is no concrete proof in the literature that RWA under shared-path protection is NP-complete. We prove that RWA under shared-path protection is NP-complete, even for $W = 1$. That RWA under sub-path protection is NP-complete follows since the former is a special case of the latter when the number of domains is one. We formally state the decision version of the RWA under shared-path protection (RWASPP) as follows.

Instance: A graph G with W wavelengths per link, and a set of lightpath requests $\mathcal{D}$.

Question: Does there exist a set of $|\mathcal{D}|$ link-disjoint path-pairs, one for each lightpath request, satisfying shared-path-protection constraints?

THEOREM 3.1 *RWA under shared-path protection (RWASPP) is NP-complete.*

Proof: *Please see Appendix 3.A.* ■

A direct application of Theorem 3.1 is that RWA under shared or sub-path protection is NP-complete for wavelength-convertible networks since the proof holds when the number of wavelengths is unity.

3.3 ILP Formulation for RWA with Sub-Path Protection

We formulate the RWA problem under sub-path protection for a given set of lightpath requests under a non-blocking model, i.e., all lightpath requests can be successfully routed. The formulation turns out to be an integer linear program (ILP). As combined RWA ILP [Ou et al., avis] is too time-and-space intensive due to the NP-completeness of the problem, we develop a two-step ILP for routing and for wavelength assignment separately. There are two main reasons for developing a two-step ILP for separated routing and wavelength assignment. First, pure routing without wavelength assignment becomes much more computationally efficient because the stringent wavelength-continuity constraint is gone. Meanwhile, we can still take into consideration backup sharing in the pure routing step, as we will show later in the formulation. Second, after routing has been done, wavelength assignment can be performed on a per-domain basis due to the wavelength-conversion capability of DBNs. The complexity of wavelength assignment is significantly reduced because each domain has a lot fewer lightpath requests to consider: Each domain only needs to worry about its intra-domain lightpath requests and a portion of the inter-domain lightpath requests which traverse this domain (compared to the entire set of lightpath requests in the no-domain case). We remark that both the routing problem and the wavelength-assignment problem are still NP-complete.

We first find, for each lightpath request, two fiber-disjoint paths with respect to sub-path-protection constraints while maximizing the potential sharing between backup paths; then, we convert each inter-domain lightpath to a collection of intra-domain lightpaths by partitioning its inter-domain fiber-disjoint paths at the common DBNs they traverse; finally, we assign wavelengths to all the sub-paths within one domain, and repeat the wavelength-assignment process for each domain.

Below we first define the notations; then, we present the split (two-step) ILP formulation; finally, we demonstrate the equivalence of the ILP to the original problem. We assume that DBNs are wavelength convertible while other nodes are wavelength continuous in these example formulations shown below, but they can be generalized.

3.3.1 Notations

The following are given as inputs:

- N: number of nodes in the network.
- W: number of wavelengths available on each link. We use w to index each wavelength, i.e., $1 \leq w \leq W$. For the formulations reported here, we assume that the same number of wavelengths are available on all links, but this too can be generalized.
- $E = \{ij\}$: set of unidirectional links in the network. We use ij to denote the link between node i and node j, and (i, j) to denote node pair i and j.
- $\mathcal{D}$: set of lightpath requests, $|\mathcal{D}| = D$. We use k to denote the kth lightpath request, and we use k_s and k_d to denote the source and the destination nodes of this lightpath request.
- $\mathcal{P}$: partitioning configuration.
- $\mathcal{R}$: set of domain-border nodes.
- C_{ij}: cost associated with link ij.

3.3.2 Sub-Path Protection: Split ILP Formulation

3.3.2.1 Part I: Routing

Part I computes the primary and backup paths for each lightpath request subject to fiber-disjoint-path-pair constraint C.1 and domain constraints C.5-C.6 (in Section 3.2.4), taking into consideration the maximal wavelength-link sharing potential between backup paths.

Part I solves for the following variables:

- F_{ij}^k takes on the value 1 if the primary path for lightpath k utilizes link ij; 0 otherwise.
- S_{pq}^k takes on the value 1 if the backup path for lightpath k utilizes link pq; 0 otherwise.
- $\delta_{pq}^{k,ij}$ takes on the value 1 if lightpath k utilizes link ij along its primary path and link pq along its backup path.
- W_{ij} is the total number of wavelengths required on link ij.

Objective: The objective function (3.1) below minimizes the total number of resources required. If C_{ij} is 1 for all the links, then the objective is to minimize the total number of wavelength-links (as in [Ramamurthy et al., 2003]); if C_{ij} is the mileage of link ij, then the objective is to minimize the total wavelength miles (as in [Fumagalli et al., 1999]), which effectively minimizes the average network delay since the link propagation delay dominates the processing delay at intermediate nodes as long as link utilization is not too close to unity [Mukherjee et al., 1996]. C_{ij} can use complex cost functions as in [Van Caenegem et al., 1998, Miyao and Saito, 1998] as long as it is linear. For the numerical results reported later, we let C_{ij} to be 1 to minimize the total number of wavelength-links. In general, the objective function is:

$$Minimize \qquad \sum_{ij \in E} (W_{ij} \times C_{ij}) \tag{3.1}$$

Constraints:

The first set of constraints, Eqn. (3.2) below, is the multicommodity-flow-conservation equations that account for the routing of the primary path for each lightpath request:

$$\sum_j F_{ij}^k - \sum_j F_{ji}^k = \begin{cases} 1 & \text{if } i = k_s \\ -1 & \text{if } i = k_d \\ 0 & \text{otherwise} \end{cases} \tag{3.2}$$
$$\forall\, 1 \leq k \leq D, 1 \leq i \leq N$$

Similarly, the set of constraints, Eqn. (3.3) below, is the multicommodity-flow-conservation equations that account for the routing of the backup path for a lightpath when a link fails. We remark that $\delta_{pq}^{k,ij}$ and S_{ij}^k are closely related and Eqn. (3.3) can be implemented by using S_{ij}^k too. The reason we introduce $\delta_{pq}^{k,ij}$ is to maximize backup sharing. These constraints are:

$$\sum_q \delta_{pq}^{k,ij} - \sum_q \delta_{qp}^{k,ij} = \begin{cases} F_{ij}^k & \text{if } p = k_s \\ -F_{ij}^k & \text{if } p = k_d \\ 0 & \text{otherwise} \end{cases} \tag{3.3}$$
$$\forall\, 1 \leq k \leq D,\ ij \in E,\ 1 \leq p \leq N$$

The set of constraints in Eqns. (3.4)-(3.4) derives S_{pq}^k from $\delta_{pq}^{k,ij}$:

$$S_{pq}^k \leq \sum_{ij \in E \wedge i,j \in A} \delta_{pq}^{k,ij} |E| \times S_{pq}^k \geq \sum_{ij \in E \wedge i,j \in A} \delta_{pq}^{k,ij} \tag{3.4}$$
$$\forall\, k,\ A \in \mathcal{P},\ pq \in E \wedge p,q \in A$$

The set of constraints in Eqn. (3.5) ensures that the primary and backup paths of one lightpath are fiber-disjoint:

$$\delta_{ij}^{k,ij} = 0,\ \delta_{ij}^{k,ji} = 0 \quad \forall\, k, \forall ij, ji \in E \tag{3.5}$$

The set of constraints in Eqns. (3.6)-(3.7) states the capacity requirements on each link. Please note that backup sharing is implicitly captured in Eqn. (3.6). The limit on the number of wavelengths per fiber is enforced by Eqn. (3.7).

$$\begin{aligned} \sum_k (F_{ij}^k + \delta_{ij}^{k,pq}) &\leq W_{ij} \quad (3.6) \\ \forall\, A \in \mathcal{P}, ij \in E \wedge i,j &\in A,\ pq \in E \wedge p,q \in A \\ W_{ij} &\leq W \quad \forall\, ij \in E \quad (3.7) \end{aligned}$$

The set of constraints in Eqns. (3.8)-(3.9) is domain related, and when combined with Eqn. (3.6) they ensure that domains are autonomous. Equation (3.8) applies to intra-domain lightpath requests, i.e., intra-domain lightpath requests should not use resources of other domains[6]. If the predicate ($\forall\, k \wedge (\exists A \in \mathcal{P}$ s.t. $k_s, k_d \in A)$) is true, then the source node, k_s, and the destination node, k_d, belong to the same domain A. (We remark that k_s and k_d may be DBNs and thus they can belong to multiple domains. If this is the case, we randomly pick one domain that they belong to.) Equation (3.9) applies to inter-domain lightpaths, and it ensures that primary and backup paths of an inter-domain lightpath exit a domain (and enter another domain) through a common DBN. The predicate ($\forall\, A' \in \mathcal{P}, \forall\, k \wedge \neg(k_s, k_d \in A')$) is true if and only if the source node, k_s, and the destination node, k_d, belong to different domains. This case may be tricky as shown in Fig. 3.1. The backup sub-path $\langle 3,6,5,7 \rangle$ traverses DBN 6 first and then exits Domain 1 via a different DBN 7. By considering the net flow of outgoing lightpath requests to other domains at DBNs, Equation (3.9) ensures that the primary sub-path $\langle 3,4,7 \rangle$ exits Domain 1 via DBN 7 instead of DBN 6. These constraints are:

$$\begin{aligned} F_{ij}^k = 0 \quad S_{ij}^k = 0 \quad (3.8) \\ \forall\, k \wedge (\exists A \in \mathcal{P} \text{ s.t. } k_s, k_d \in A), \forall\, i \notin A \vee j \notin A \end{aligned}$$

$$\sum_{\forall i \in A: ij \in E} F_{ij}^k - \sum_{\forall e \in A: je \in E} F_{je}^k = \sum_{\forall i \in A: ij \in E} S_{ij}^k - \sum_{\forall e \in A: je \in E} S_{je}^k \quad (3.9)$$
$$(\forall\, A' \in \mathcal{P}, \forall\, k \wedge \neg(k_s, k_d \in A')), \forall\, A \in \mathcal{P}, \forall\, j \in \mathcal{R} \wedge j \in A$$

[6] In our implementation of the ILP for the topology shown in Fig. 3.1, we completely got rid of these zero variables stated in Eqn. (3.8) and sped up the solution process. While it may be possible to eliminate most of these zero variables for a particular instance of the problem, note that some of them may be non-zero, and our objective is to also provide a compact mathematical formulation of the problem (which can be customized as needed for specific implementations, as we have done).

The final set of constraints states that F_{ij}^k, S_{pq}^k, and $\delta_{pq}^{k,ij}$ are binaries while W_{ij} is integer (W_{ij} is bounded by W in Eqn. (3.7)):

$$F_{ij}^k, S_{pq}^k, \delta_{pq}^{k,ij} \in \{0,1\},\ W_{ij} \in integer \qquad (3.10)$$
$$\forall\, 1 \le k \le D, ij, pq \in E$$

We remark that Part I can accommodate other objective functions such as minimizing the total number of wavelengths. To minimize the total number of wavelengths, we can use $Minimize\ \ W$ as the objective function (W becomes a variable in that case). The flow-conservation constraints (3.2) and (3.3) may not be efficient in avoiding loops in the paths. We can add additional constraints as proposed in [Nucci et al., 2002] to address this problem. The basic idea is to ensure that a path only traverses fibers that are part of a subset of the physical topology called a covering tree. To model the routing problem under shared-path protection, we simply let the number of domains, $|\mathcal{P}|$, be unity and the DBN set, $\mathcal{R}$, be empty.

Based on the routing results, we segment the fiber-disjoint path pair of an inter-domain lightpath request according to the common DBNs the path pair traverses; we then collect all the fiber-disjoint path pairs of a domain and use the following Part II to assign wavelength to every path on a per-domain basis.

One critical assumption here, and later for the heuristic, is that all the DBNs are wavelength convertible. Without this assumption, wavelength assignment for multiple domains is coupled by the wavelength-continuity constraint and per-domain independent wavelength assignment may not be possible. In practice, optical domains are expected to be isolated by transponders [Strand and Xue, 2001], making wavelength conversion available at DBNs and making this assumption realistic.

3.3.2.2 Part II: Wavelength Assignment

Part II assigns wavelengths to the primary and backup paths of all the lightpaths within one domain subject to shared-path-protection constraints C.2-C.4 (in Section 3.2.4) and wavelength-continuity constraints. The following are given as inputs:

- p_p^k is the primary path for lightpath k. We use $ij \in p_p^k$ to denote that p_p^k traverses link ij.
- p_b^k is the backup path for lightpath k.
- w_{ij} is the number of wavelengths used by primary paths on link ij. (We can derive w_{ij} from p_p^k, i.e., $w_{ij} = \sum_{k:ij \in p_p^k} 1$, since primary paths are fixed.)

Part II solves for the following variables:

- F_w^k takes on the value 1 if the primary path for lightpath k utilizes wavelength w.
- S_w^k takes on the value 1 if the backup path for lightpath k utilizes wavelength w.
- m_{ij}^w takes on the value 1 if wavelength w on link ij is used by some backup path.
- s_{ij} denotes the number of wavelengths used by backup paths on link ij.

Objective: The objective function (3.11) minimizes the total amount of resources used by backup paths since the amount of resources used by primary paths is the fixed value, $\sum_{ij \in E}(w_{ij} \times C_{ij})$:

$$\text{Minimize} \qquad \sum_{ij}(s_{ij} \times C_{ij}) \tag{3.11}$$

The set of constraints in Eqns. (3.12)-(3.13) ensures that each lightpath is assigned a wavelength:

$$\sum_{w=1}^{W} F_w^k \;=\; 1 \quad \forall\, k \tag{3.12}$$

$$\sum_{w=1}^{W} S_w^k \;=\; 1 \quad \forall\, k \tag{3.13}$$

The set of constraints in Eqns. (3.14)-(3.15) decides whether a wavelength on a link is used by some backup path:

$$m_{ij}^w \leq \sum_{k:ij \in p_b^k} S_w^k \quad \forall ij \in E \tag{3.14}$$

$$D \times m_{ij}^w \geq \sum_{k:ij \in p_b^k} S_w^k \quad \forall ij \in E \tag{3.15}$$

The set of constraints in Eqn. (3.16) states that the primary and backup paths can not share any wavelength-link:

$$m_{ij}^w + \sum_{k:ij \in p_p^k} F_w^k \leq 1 \quad \forall ij \in E,\; w \tag{3.16}$$

The set of constraints in Eqn. (3.17) ensures that two backup paths can not share any wavelength-link if their corresponding primary paths are not fiber-disjoint:

$$\sum_{k:ij \in p_p^k \wedge pq \in p_b^k} S_w^k \leq 1 \quad \forall w,\; \forall ij, pq \in E \tag{3.17}$$

The set of constraints in Eqn. (3.18) decides the number of wavelengths used by backup paths on link ij:

$$s_{ij} = \sum_{w=1}^{W} m_{ij}^{w} \quad \forall ij \in E \tag{3.18}$$

The set of constraints in Eqn. (3.19) ensures that the number of wavelengths used on each link is no more than the total number of wavelengths available:

$$w_{ij} + s_{ij} \leq W \quad \forall ij \in E \tag{3.19}$$

The final set of constraints in Eqn. (3.20) states that F_w^k, S_w^k, m_{ij}^w are binaries and s_{ij} is integer (s_{ij} is bounded by W in Eqn. (3.19)):

$$\begin{gathered} F_w^k, S_w^k, m_{ij}^w \in \{0,1\}, \ s_{ij} \in integer \\ \forall\, 1 \leq k \leq D, 1 \leq w \leq W, ij \in E \end{gathered} \tag{3.20}$$

Note that we can also use *Minimize W* as the objective to minimize the number of wavelengths.

3.3.3 Equivalence of the Split ILP and the Original Problem

We demonstrate the equivalence of the split ILP and the RWA problem with sub-path protection by showing that any valid solution produced by the ILP satisfies the original Constraints C.1-C.6 and vice versa. First, the primary and backup paths for a lightpath request are formed by the flow-conservation constraints in Eqns. (3.2) and (3.3). Constraint C.1 is ensured in Eqn. (3.5). Constraints C.2 and C.3 are formulated in Eqns. (3.12), (3.13), and (3.16). Constraint C.4 is enforced by Eqn. (3.17). Constraint C.5 is ensured in Eqn. (3.8). Constraint C.6 is captured in Eqn. (3.9). The constraint on the number of wavelengths per fiber is accommodated by Eqns. (3.6), (3.7), and (3.19). The objective of the problem to minimize total network cost is formulated in the two objective functions (3.1) and (3.11). The other equations, such as Eqns. (3.4)-(3.4), are used to derive intermediate variables. Later, in Section 3.5, we shall present illustrative results from the split ILP.

3.4 Heuristic

While ILPs are useful in providing insights into the nature of the problem, they may be hard to solve for large networks with dozens of wavelengths per link because of the long computational time caused by the NP-completeness of the original problem (please refer to Section 3.3). We propose a heuristic for RWA under sub-path-protection constraints for a given set of lightpath requests and evaluate its effectiveness. The crucial observation from the split

ILP is that, once fiber-disjoint paths are found for each inter-domain lightpath request, wavelength assignment is of local (a domain) matter. This inspires us to investigate algorithms to segment inter-domain lightpaths, and then to assign wavelengths on a per-domain basis.

The heuristic has three phases. Phase 1 computes the shortest fiber-disjoint path pair for each lightpath request subject to domain constraints and then partitions each inter-domain lightpath into a set of sub-paths according to the common DBNs the lightpath's primary and backup paths traverse. The solution found in this phase serves as the seed solution for later optimization. Phases 2 and 3 work on a per-domain basis. Phase 2 assigns wavelengths to all the paths of a domain. Phase 3 first reroutes—for all the fiber-disjoint path pairs of a domain—the backup paths to increase backup sharing with respect to shared-path-protection constraints, and then it rearranges the primary paths to reduce the number of wavelength-links a primary path uses.

The heuristic takes as input a network as a weighted, directed graph G; the number of wavelength, W, on each fiber; a proper partitioning configuration $\mathcal{P}$; and a given set of lightpath requests $\mathcal{D}$. The heuristic outputs the routing and wavelength assignment of the primary and backup paths for each lightpath request.

3.4.1 Phase 1: Find Shortest Path Pair for Each Lightpath with Respect to Domain Constraints

Phase 1 computes the shortest fiber-disjoint paths for each lightpath request subject to domain constraints, i.e., the primary and backup paths of an inter-domain lightpath exit a domain (and enter another domain) through a common DBN and an intra-domain lightpath cannot use resources of other domains. As the routing problem under shared-path (or sub-path) protection constraints with multiple lightpath requests is NP-complete, Phase 1 works on a per-lightpath basis (backup sharing will be considered later).

For any lightpath request $k \in \mathcal{D}$, if k is an intra-domain lightpath request, the heuristic directly applies a known shortest link-disjoint path-pair algorithm (e.g., Suurballe's algorithm [Suurballe and Tarjan, 1984] or the algorithm in [Bhandari, 1999]); if k is an inter-domain lightpath request, the heuristic constructs a weighted, directed auxiliary graph G_a^k such that, when transformed back to G, the shortest path between (k_s, k_d) in G_a^k is the shortest fiber-disjoint path pair for (k_s, k_d) in G. G_a^k has k_s, k_d, and all the DBNs as its vertices. The cost of a link in G_a^k is the cost of the shortest fiber-disjoint path pair between the end-nodes of that link in G. Phase 1 works as follows:

1 Construct a weighted, directed graph $G_a = (V_a, E_a)$ where:

1.1 $V_a = \mathcal{R}$, the set of DBNs.

1.2 $\forall\, i \neq j \in V_a, ij \in E_a$ if and only if $\exists A \in \mathcal{P}$ s.t. $i, j \in A$.

1.3 C_{ij}, the cost of link ij, is the cost of the shortest fiber-disjoint path pair between node pair (i, j) among all such path pairs in any domain A such that $i \in A \wedge j \in A$ in G.

Then, the heuristic applies the following steps to each inter-domain lightpath k (k is an inter-domain lightpath if and only if $\forall\ A \in \mathcal{P}, \neg(k_s, k_d \in A)$).

2 Construct a weighted, directed graph $G_a^k = (V_a^k, E_a^k)$ as follows (assume $k_s \in A_s, k_d \in A_d, A_s \neq A_d \in \mathcal{P}$):

2.1 $V_a^k = V_a \cup \{k_s, k_d\}$.

2.2 E_a^k comprises the following three components.

2.2.1 $\forall r \in \mathcal{R} \wedge r \in A_s, k_s r \in E_a^k, C_{sr}$ is the cost of the shortest fiber-disjoint path pair between (k_s, r) in A_s.

2.2.2 $\forall r \in \mathcal{R} \wedge r \in A_d, r k_d \in E_a^k, C_{rd}$ is the cost of the shortest fiber-disjoint path pair between (r, k_d) in A_d.

2.2.3 $\forall r_1 \neq r_2 \in \mathcal{R} \wedge \neg(r_1, r_2 \in A_s \vee r_1, r_2 \in A_d), r_1 r_2 \in E_a^k$, and it has the same cost as $r_1 r_2$ in G_a.

3 Compute the shortest path between (k_s, k_d) in G_a^k. Denote the shortest path by $\langle r_0 = k_s, r_1, \cdots, r_k, r_{k+1} = k_d \rangle$.

4 Concatenate the shortest fiber-disjoint path pair between r_i and r_{i+1} in G, for $i = 0, 1, \cdots, k$, to obtain the shortest fiber-disjoint path pair for lightpath k in G.

Figure 3.3 shows G_a and $G_a^{3,16}$ for the graph in Fig. 3.1. Conceptually, we only need G_a^k. The reason we have G_a is to avoid unnecessary repetitive computation in Step 2.2.3.

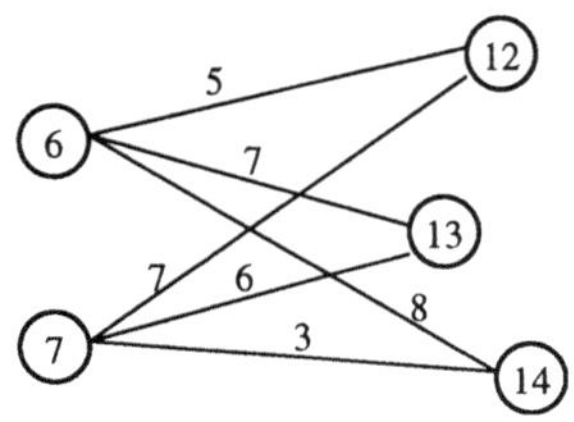

(a) G_a

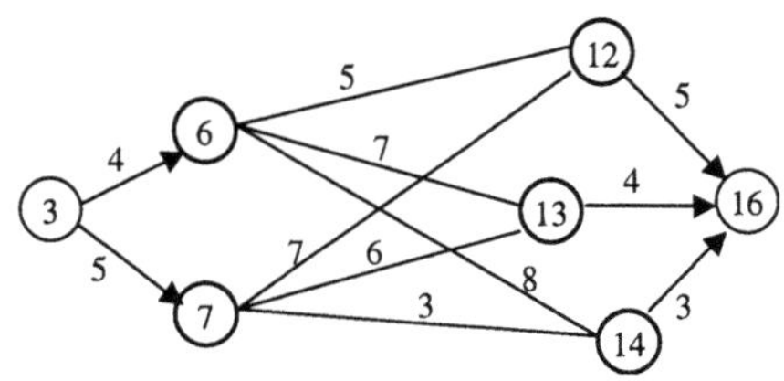

(b) $G_a^{3,16}$

Figure 3.3. Auxiliary graphs for the network shown in Fig. 3.1. The number on a link is the cost of that link, assuming the cost of each link in Fig. 3.1 is unity. A link with arrow is a unidirectional link; a link without an arrow is a bidirectional link.

3.4.2 Phase 2: Wavelength Assignment

Wavelength-assignment heuristics have been extensively studied [Zang et al., 2000]. We use the wavelength-assignment algorithm in [Sahin and Azizoglu, 2001] for the sake of simplicity because our main interest is a heuristic that finds efficient solutions. The basic ideas are to transform the wavelength assignment of primary and backup paths to the graph-coloring problem, and to employ a sequential coloring [Banerjee and Mukherjee, 1996, Matula et al., 1972] algorithm. Phase 2 works on a per-domain basis since an inter-domain lightpath is segmented into a set of sub-paths in Phase 1. The shorter path of the fiber-disjoint path pair is considered as the primary path, and the longer one is the backup path. Phase 2 works as follows.

First, we construct an undirected graph, G_c, in which each path is treated as a vertex. The edges are constructed as follows: two primary paths are adjacent in G_c if they traverse a common fiber; a primary path and a backup path are adjacent if they traverse a common fiber; two backup paths are adjacent if their corresponding primary paths traverse a common fiber. We then sequentially color all the vertices such that adjacent vertices have different colors. The sequential order is smallest-last [Matula et al., 1972] in which the vertex of smallest degree is colored last.

3.4.3 Phase 3: Optimization

The purpose of this optimization procedure is to rearrange the primary and backup paths to reduce the total number of wavelength links used. A similar approach can be found in [Doshi et al., 1999]. In step one of this two-step procedure, we reroute the backup paths to increase backup sharing. In step two, we rearrange the primary paths since there may be a shorter path available after rerouting the backup paths. Please note that Phase 3 also works on a per-domain basis.

Before describing the optimization procedure, we prove that the problem of routing the backup paths to find the optimal solution is NP-complete. We formally state the optimal-backup-routing (OBR) problem as follows:

OBR: Given a WDM network with W wavelengths on each link as a weighted, directed graph $G = (V, E, C)$, an integer c, and a set $\mathcal{D}$ of lightpath requests whose primary paths, p_p^k $(1 \leq k \leq |\mathcal{D}|)$, are known, does there exist a backup path for each lightpath request subject to shared-path-protection constraints (C.1-C.4 in Section 3.2.4) such that the total cost of the backup paths is no more than c?

THEOREM 3.2 *The OBR problem is NP-complete.*

Proof: Please see Appendix 3.B. ■

Step one starts with randomly picking one lightpath, say k, with primary path p_p^k and backup path p_b^k. Next, remove the backup path p_b^k and update the cost of wavelength w on link ij, C_{ij}^w, as follows ($+\infty$ is a large number, e.g., the diameter of the network times the maximum link cost, where the diameter of the network is defined as the total number of hops of the longest path in the network):

$$C_{ij}^w = \begin{cases} +\infty & \text{if } p_p^k \text{ traverses link } ij\text{, or wavelength } w \text{ on} \\ & \text{link } ij \text{ is used by some primary paths, or} \\ & \text{wavelength } w \text{ on link } ij \text{ is used by some} \\ & \text{backup path whose primary path is not} \\ & \text{mutually diverse with } p_p^k \\ C_{ij} & \text{if wavelength } w \text{ on link } ij \text{ is not used} \\ 0 & \text{otherwise} \end{cases}$$

Then, compute the minimal-cost path from k_s to k_d on each wavelength layer, and pick the one with the minimal cost as the backup path. Note that the fiber-disjoint constraints and the constraint that two backup paths cannot share resources if their primary paths are not fiber-disjoint are ensured by the cost function, and the newly computed backup has a cost no larger than p_b^k.

Lastly, repeat this process for a predefined number of times or until it converges, i.e., the new backup path does not have smaller cost compared to the previous one for a predefined number of times.

Instead of simply admitting the newly computed lightpath with minimum cost, we tried a simulated-annealing based approach. In this approach, we admit the new minimum-cost lightpath only when the cost difference, $\triangle$, between the old backup path and the new one is greater than a threshold, which decreases gradually; or $e^{-\triangle}$ is less than a random number between 0 and 1. We found that simply admitting the newly computed lightpath works better in terms of resource utilization when the number of iterations is modest. The reason is that the simulated-annealing-based approach needs more time to converge.

Step two is similar to Step one, except that we remove the primary path p_p^k and update the cost of wavelength w on link ij according to the following cost function:

$$C_{ij}^w = \begin{cases} +\infty & \text{if } p_b^k \text{ traverses link } ij\text{, or } p_b^k \text{ shares some} \\ & \text{wavelength-link with some backup path whose} \\ & \text{primary path traverse link } ij\text{, or wavelength } w \\ & \text{on link } ij \text{ is used by some lightpath} \\ C_{ij} & \text{otherwise} \end{cases}$$

We prove the correctness of the heuristic as follows. Let $H(G, W, \mathcal{P}, \mathcal{D})$ be the solution given by the heuristic, and let $H_i(G, W, \mathcal{P}, \mathcal{D})$ be the intermediate result given by Phase i ($i = 1, 2, 3$). We first prove that $H_1(G, W, \mathcal{P}, \mathcal{D})$ is correct with respect to routing under domain constraints and fiber-disjoint constraints; then we prove that $H_2(G, W, \mathcal{P}, \mathcal{D})$ is correct with respect to routing and wavelength assignment under sub-path-protection constraints (C.1-C.6 in Section 3.2.4), and $H_3(G, W, \mathcal{P}, \mathcal{D})$ maintains correctness.

CLAIM 1 *The path pair (p_p^k, p_b^k) for lightpath k given by $H_1(G, W, \mathcal{P}, \mathcal{D})$ satisfies the following:*

1 *p_p^k and p_b^k are fiber-disjoint.*

2 *p_p^k and p_b^k do not use resources of other domains if k is an intra-domain lightpath; p_p^k and p_b^k exit one domain (and enter another domain) through a common DBN if k is an inter-domain lightpath.*

3 *The total cost of p_p^k and p_b^k is minimal among all such pairs.*

Proof: If k is an intra-domain lightpath, then Conditions 1 and 3 follow directly from the proof given in [Suurballe and Tarjan, 1984], and Condition 2 is true because the heuristic is carried out in the domain k_s and k_d belong to. Hereafter, we assume that k is an inter-domain lightpath request, and we let p_s^k be the shortest path found by $H_1(G, W, \mathcal{P}, \mathcal{D})$ in G_a^k. We prove the claim for an inter-domain lightpath as follows.

1 Suppose p_p^k and p_b^k are not fiber-disjoint. By the construction of G_a^k, p_s^k must use at least two links (in G_a^k) whose corresponding fiber-disjoint path pairs in G are in the same domain because domains do not share any link (please note that any link p_s^k traverses in G_a^k corresponds to two fiber-disjoint paths in G). Let $\langle r_i, r_j\rangle$ and $\langle r_x, r_y\rangle$ be two such links. Then r_i, r_j, r_x, r_y must be in the same domain; otherwise, their corresponding fiber-disjoint path pairs will not be in the same domain. Without loss of generality, assume that p_s^k traverses link $\langle r_x, r_y\rangle$ after $\langle r_i, r_j\rangle$, i.e., p_s^k is of the form $\langle k_s, \cdots, r_i, r_j, \cdots, r_x, r_y, \cdots, k_d\rangle$. By the construction of G_a^k, the link $\langle r_i, r_y\rangle$ has less cost than the path $\langle r_i, r_j, \cdots, r_x, r_y\rangle$; thus, $\langle k_s, \cdots, r_i, r_y, \cdots, k_d\rangle$ has less cost than p_s^k. This contradicts that p_s^k is the shortest path. Thus, p_p^k and p_b^k are fiber-disjoint.

2 By the construction of G_a^k, each link along p_s^k corresponds to a fiber-disjoint path pair in G, and the path pair (p_p^k, p_b^k) is formed by concatenating all such fiber-disjoint path pairs. That p_p^k and p_b^k exit one domain (and enter another domain) through a common DBN follows since V_a^k includes only DBNs, k_s, and k_d.

3 This condition holds because segments of a shortest path are shortest sub-paths [Cormen et al., 1990], and the fiber-disjoint path pair found in a domain using Suurballe's algorithm is the shortest among all such pairs in that domain [Suurballe and Tarjan, 1984].

■

The solution given by $H_2(G, W, \mathcal{P}, \mathcal{D})$ satisfies domain constraints (C.5-C.6 in Section 3.2.4) because Phase 2 works on a per-domain basis. The solution given by $H_2(G, W, \mathcal{P}, \mathcal{D})$ satisfying shared-path-protection constraints (C.1-C.4 in Section 3.2.4) follows from the construction of G_c (defined in Section 3.4.2) and the correctness of the sequential coloring algorithm. The solution given by $H_3(G, W, \mathcal{P}, \mathcal{D})$ maintaining correctness follows from the definition of the wavelength-link cost function and the correctness of shortest-path algorithm [Cormen et al., 1990].

3.4.4 Complexity

The complexity of Phase 1 is $O(N^4 + D \cdot N^3)$ (assuming Dijkstra's algorithm for shortest-path computation). The complexity of Step 1.1-Step 1.3 is $O(N^4)$ since $V_a = |\mathcal{R}| \leq N, E_a < |V_a|^2 <= N^2$, and each link of G_a takes time $O(N^2)$ to compute. The complexity of Step 2.1 is $O(N)$, and Step 2.2 is $O(N^3)$ (Step 2.2.1: $O(N^3)$, Step 2.2.2: $O(N^3)$, and Step 2.2.3: $O(N^2)$). Thus, the complexity of Step 2 is $O(D \cdot N^3)$. The complexities of Step 3 and Step 4 are $O(N^2)$ and $O(A)$, respectively.

The complexity of Phase 2 is $O(N \cdot D^2)$. The complexity of constructing G_c is $O(N \cdot D^2)$ because we need to compare $2 \cdot D$ paths, each of which is of length less than N. The complexity of sequential coloring is $O(D \cdot W)$, where W is the maximum number of colors and W is no more than $2 \cdot D$.

The complexity of Phase 3 is $O(K \cdot W \cdot N^3 \cdot D)$, where K is the number of times Step 1 (in Phase 3) repeats. The complexity of calculating the cost function for each wavelength-link on one wavelength layer is $O(N \cdot D)$. The complexity of the shortest-path algorithm is $O(N^2)$. The complexity of computing the shortest path on one wavelength layer (including calculating the cost function) is $O(N^3 \cdot D)$ since each wavelength layer can have $O(N^2)$ links. Since we have W wavelength layers and we need to repeat the entire process K times, the complexity of Phase 3 is $O(K \cdot W \cdot N^3 \cdot D)$.

The total complexity of the heuristic is thus $O(N^4 + D \cdot N^3 + |\mathcal{P}|(N \cdot D^2 + \cdot K \cdot W \cdot N^3 \cdot D)$=$O(|\mathcal{P}| \cdot K \cdot N^3 \cdot D^2)$ since, in general, D is larger than N and W. We remark that this is a loose worst-case upper bound, and the complexity is much less in practice since a domain only has $\lceil \frac{N}{|\mathcal{P}|} \rceil$ nodes on average (while the complexity analysis considers that a domain has N nodes). Furthermore, Phase 1 can be carried out in parallel for each lightpath request, and Phase 3 can be carried out in parallel on each wavelength layer.

3.5 Results and Discussions

We compare the recovery time, survivability, scalability, and resource utilization of sub-path protection with those of shared-path protection by applying the split ILP and the heuristic to the network in Fig. 3.1. (We remark that shared-link protection is much less resource efficient than shared-path protection [Ramamurthy et al., 2003, Iraschko et al., 1998]; thus, shared-link protection is not compared to sub-path protection in this study.) Shared-path protection is a special case of sub-path protection when the number of domains is one, and the results for shared-path protection are obtained by treating the entire network as one domain.

Please note that the fact that the DBNs are wavelength convertible under sub-path protection while the corresponding nodes are wavelength continuous under shared-path protection does not bias the results much *under a non-blocking model.* While wavelength-conversion capability at DBNs is critical to simplify algorithm design and to achieve management convenience for sub-path protection, wavelength-conversion capability at the corresponding nodes plays a much less important role for shared-path protection under a non-blocking model. The underlying assumption for a non-blocking model is that all the lightpath requests can be satisfied. This implies that sufficient number of wavelengths exists. As a result, the need for wavelength converters is small. For example, if a lightpath request needs to traverse a wavelength-convertible node to convert from wavelength λ_1 to λ_2, chances are that this lightpath request may be routed using a different wavelength, say λ_6. However, it is valid that the existence of a few wavelength converters may slightly reduce the required number of wavelengths for shared-path protection.

3.5.1 Recovery Time

The recovery time[7] of link $\langle i, j\rangle$ with respect to a lightpath between node-pair (s, d), T_{ij}^{sd}, is defined as the time period over which data on the (s, d) lightpath is lost due to the failure of link $\langle i, j\rangle$. We use the following notations to derive the general formula of T_{ij}^{sd} for shared-path protection:

- d_{is}: propagation delay from node i to node s;
- h_{is}: number of hops from node i to node s;
- h_b: number of hops on the backup path between node pair (s, d);
- F: failure-detection time;
- X: crossconnect-configuration time; and
- M: message-processing time (including queueing delay) at a node.

[7]Note that the recovery time here is similar to the protection-switching time defined in [Ramamurthy et al., 2003] and the restoration speed defined in [Doshi et al., 1999], but it has some improvements, e.g., pipelining the switch configurations.

First, consider the time period d_{is} right before link $\langle i,j \rangle$ fails. Node s has been sending data, which will be lost when link $\langle i,j \rangle$ fails. Second, upon detecting a link failure, node i sends[8] a failure indication signal (FIS) to node s. This takes time $F + d_{is} + (h_{is} + 1) \times M$. Finally, upon receiving the FIS, node s sends a setup message to node d along the backup route, and waits for time $X + (h_b + 1) \times M$ before it switches to the backup route (configuring the crossconnects can be done in parallel). In summary,

$$T_{ij}^{sd} = F + 2 \times d_{is} + (h_{is} + 1) \times M + X + (h_b + 1) \times M \ .$$

The recovery time of link $\langle i,j \rangle$, T_{ij}, is the recovery time averaged over all the lightpaths that traverse $\langle i,j \rangle$, i.e., $T_{ij} = (\sum_{(s,d)\in\Gamma} T_{ij}^{sd})/|\Gamma|$, where Γ is the set of lightpaths whose primary paths traverse $\langle i,j \rangle$. The network-wide *average recovery time*, T, is defined as the summation of the probability[9] that link $\langle i,j \rangle$ fails (with probability p_{ij}) times the recovery time of link $\langle i,j \rangle$, i.e., $T = \sum_{\langle i,j \rangle \in E} p_{ij} \times T_{ij}$. The network-wide *longest recovery time* is the maximum of T_{ij}^{sd} over all possible (s,d) and $\langle i,j \rangle$ combinations.

Under sub-path protection, the recovery time is defined the same as above for an intra-domain lightpath request. For an inter-domain lightpath request, the recovery-time calculation is carried out for each segmented intra-domain sub-path.

For our numerical examples, we assume $F = 10\ \mu s$, $X = 5\ ms$, and $M = 20\ \mu s$, which appear to be consistent with current technologies. Figure 3.4 demonstrates that sub-path protection reduces both the network-wide longest recovery time and the average recovery time significantly, e.g., in this example, it reduces average recovery time by 14% (for $D = 40$) to 24% (for $D = 250$), as shown in Figs. 3.4(c) and 3.4(d), and longest recovery time by 28% (for $D = 50$) to 35% (for $D = 500$), as shown in Figs. 3.4(a) and 3.4(b). We observe from our results that the link propagation delay dominates recovery time. Thus, when the size of the network increases, the recovery time of shared-path protection will increase accordingly since the total mileage of the fiber-disjoint path pair increases. The recovery time of sub-path protection, however, will remain relatively constant since we can keep the size of each domain modest by increasing the number of domains. Even if the size of the network remains the same, the longest recovery time for shared-path protection increases gradually when the number of lightpath requests increases, as shown in Fig. 3.4(b), while the longest recovery time for sub-path protection remains relatively constant. This is because, when the number of lightpaths increases, more backup paths

[8]We assume that there is a separate packet-based control plane, which has the same topology as the network we are considering. A packet sent on the control plane follows the shortest path between the source and destination nodes.

[9]The probability that a fiber fails is proportional to the length of the fiber [Hac, 1994].

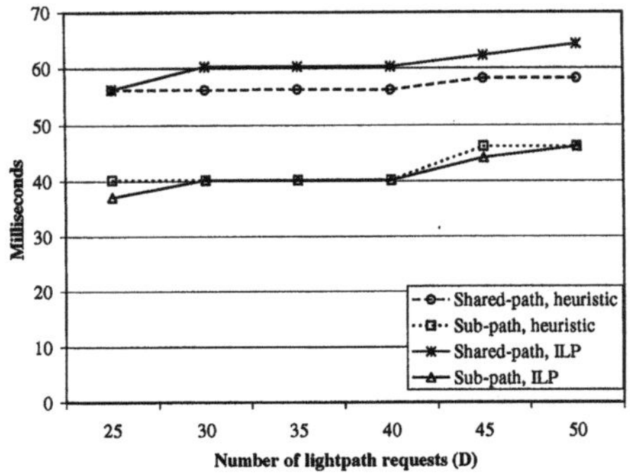

(a) Longest recovery time.

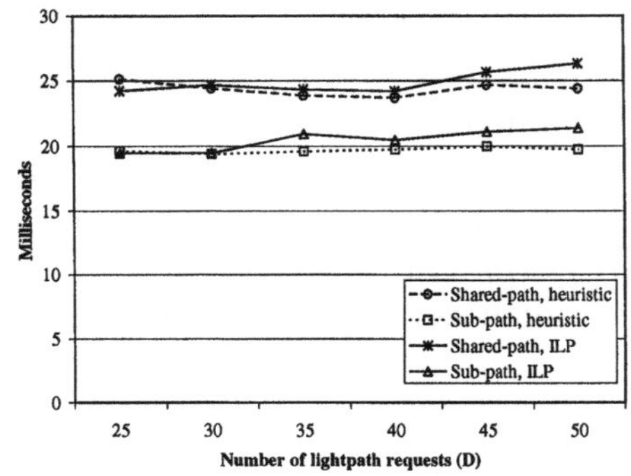

(c) Average recovery time.

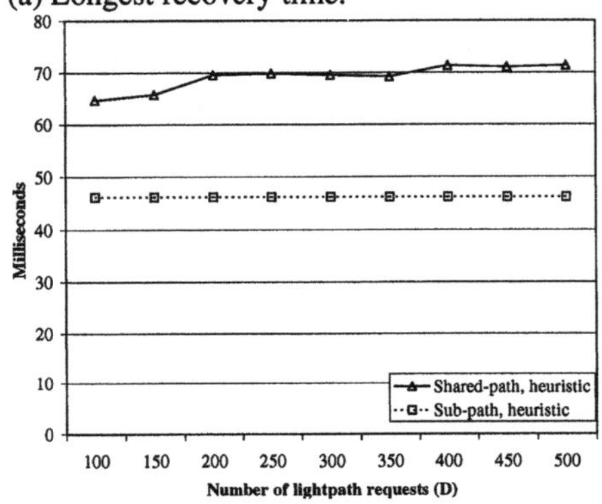

(b) Longest recovery time from heuristic.

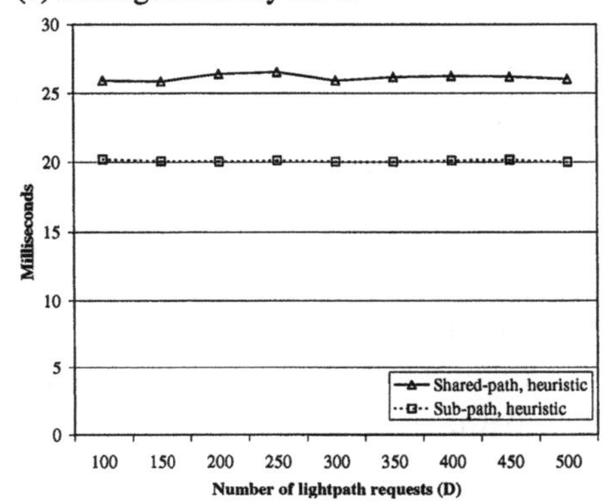

(d) Average recovery time from heuristic.

Figure 3.4. Recovery-time comparison between sub-path and shared-path protection schemes.

are likely to share wavelength links. This increased backup sharing leads to longer backup paths [Xiong et al., 2003]. Since sub-path protection has a much smaller upper bound on the domain diameter compared to the upper bound on the network diameter, it has a much smaller upper bound on recovery time. Note that, under sub-path protection, the longest recovery time is below 50 ms in all the cases. Thus, it may now be possible to guarantee the 50 ms recovery time (as SONET ring does) in large mesh networks by properly partitioning the network and applying sub-path protection.

3.5.2 Survivability

Survivability refers to the types of failures a proposed scheme can handle: link, node, or multiple failures (which [Doshi et al., 1999] refers to as restorability). Shared-path protection [Ramamurthy et al., 2003] handles single-link failures by using fiber-disjoint path pairs [Ramamurthy et al., 2003, Miyao and Saito, 1998, Kawamura et al., 1994], and single-node failure by using node-disjoint path pairs [Gerstel and Ramaswami, 2000a, Sen et al., 2001], not multiple failures. Sub-path protection combats against single-link failures the same way as shared-path protection does as shown in the split ILP. Sub-path protection can also handle non-DBN failures by finding node-disjoint path pairs and DBN failures by the approach in Section 3.2.3.

An improvement on survivability over shared-path protection is that sub-path protection can survive up to $|\mathcal{P}|$ failures as long as there is only one failure per domain because, due to the autonomous property of domains, sub-paths of the same lightpath are treated independent of one another.

3.5.3 Scalability

We consider scalability from the network control and management perspective. For network control, as pointed out in [Gerstel and Ramaswami, 2000a], any mesh protection scheme needs to maintain large distributed routing tables. However, under sub-path protection, a non-DBN node only needs to maintain state information about the domain it belongs to, and a DBN node needs to maintain state information about the domains it belongs to and the summarized connectivity information between DBNs (compared to the state information of the entire network). Signaling is another important aspect of network control. Under sub-path protection, local signaling information is filtered at DBNs and will not be disseminated to other domains. For example, sub-path protection limits fault propagation within the scope of a domain (compared to the entire network in shared-path protection) and, after a fault occurs, protection switching, the process used by the source node of an affected lightpath to signal the destination node to switch to the backup path, happens within the scope of a domain (compared to the entire network). Thus, it reduces protection-switching time (as shown in Section 3.5.1) and saves control bandwidth as well.

For network management, networks of modest size need facility management systems (FMS) to monitor network state, to localize fault, and to provide billing and other capabilities. In the absence of administrative domains, how the FMS for the entire network keeps pace with the growth of the network becomes questionable, as has been witnessed by carriers [Strand and Xue, 2001]. Under sub-path protection, each domain has its own FMS, and the function of managing the entire network is distributed to each domain. Furthermore, fault localization is limited to the scope of a domain (compared to the entire network in the absence of domains). As a result, the mean time to repair will be reduced and the overall network reliability can be improved [Akyamac et al., 2002].

In summary, domain partitioning hides information from all nodes and leads to improved scalability. This information hiding is essential to scalability, as has been justified by IP networks, which work as a collection of autonomous systems further divided into Interior Gateway Protocol (IGP) areas, such as OSPF and IS-IS areas.

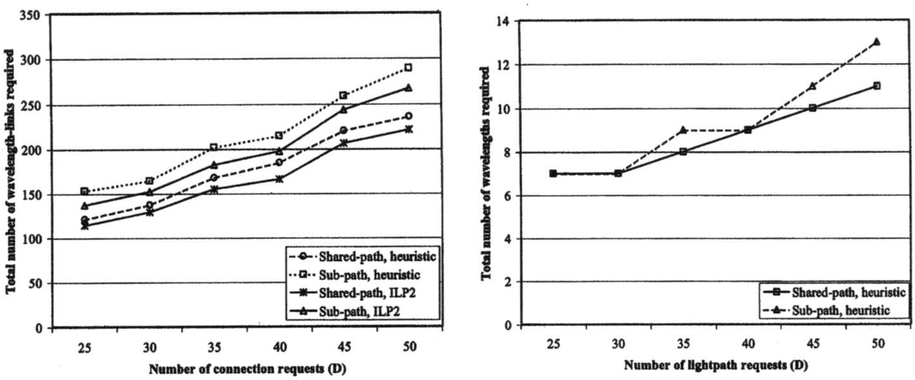

(a) (b)

Figure 3.5. Resource-utilization comparison: (a) number of wavelength-links; (b) number of wavelengths.

3.5.4 Resource Utilization

Figure 3.5(a) compares the total number of wavelength-links required for sub-path protection with that of shared-path protection by applying the split ILP and the heuristic to the network shown in Fig. 3.1 with six sets of randomly generated lightpath requests. We observe that sub-path protection requires more resources, e.g., it sacrifices network resource utilization by 17% (for $D = 35$) to 21% (for $D = 50$) from the ILP's results in this example. This is expected since sub-path protection has the domain constraints C.5-C.6 in addition to the shared-path-protection constraints C.1-C.4. The second observation is that the heuristic performs close to the split ILP for both shared-path protection and sub-path protection. Figure 3.5(b) also shows similar trend for the total number of wavelengths required. Thus, we use the heuristic to compare the resource utilization of sub-path protection with that of shared-path protection for a large number of lightpath requests.

Figure 3.6 plots the total number of wavelength-links required for sub-path protection to that of shared-path protection by applying the heuristic to the network in Fig. 3.1 with nine sets of randomly generated lightpath requests (the number of lightpaths, D, ranges from 100 to 500). While the difference between the total number of wavelength-links used by sub-path protection and shared-path protection increases as D increases, as shown in Fig. 3.6(a), the difference between the number of wavelength-links used per lightpath remains relatively constant, as shown in Fig. 3.6(b). This means that the amount of resources sacrificed due to domain constraints is proportional to the number of lightpaths. Figure 3.6(b) also illustrates the advantage of backup sharing: the

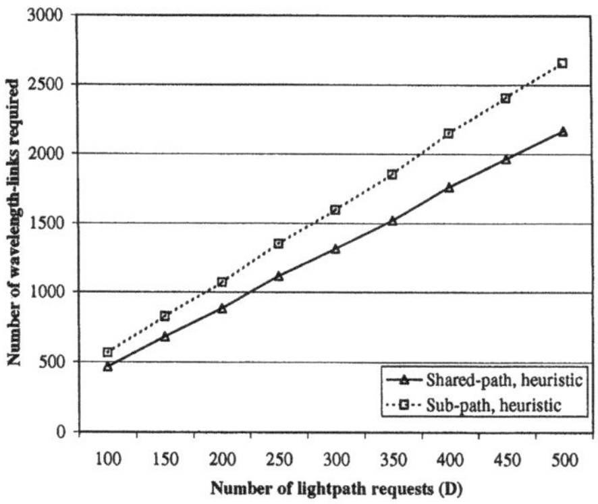

(a)

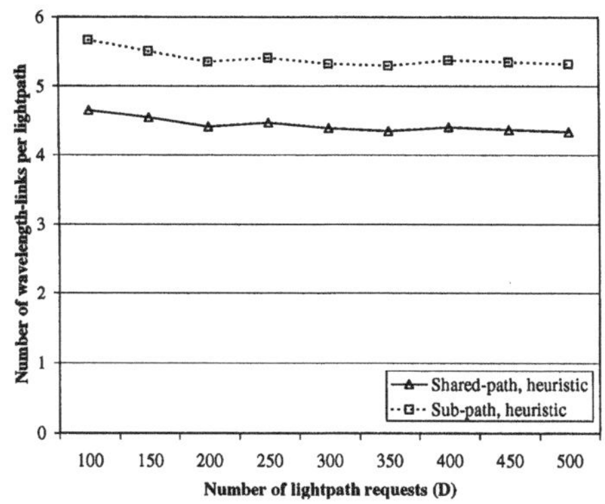

(b)

Figure 3.6. Resource-utilization comparison: number of wavelength-links used (a) for all the lightpaths and (b) per lightpath.

number of wavelength-links per lightpath in both shared-path protection and sub-path protection decreases as the number of lightpaths increases.

A counter-intuitive observation is that sub-path protection may sometimes require *less* resources than shared-path protection, as shown in Figure 3.7. The 10-node network has two wavelengths per fiber, and it needs to carry two lightpath requests between node pair $(7, 8)$ and another two lightpath requests between node pair $(0, 9)$. In both cases, the two lightpaths between node pair $(7, 8)$ will use the only two wavelengths on link $\langle 6, 9\rangle$ because there is only one fiber-disjoint path pair between node pair $(7, 8)$ ($\langle 7, 8\rangle$ and $\langle 7, 6, 9, 8\rangle$). If any of the two wavelengths on link $\langle 6, 9\rangle$ is used by a primary path between node pair $(7, 8)$, then there is no fiber-disjoint path pair for one of the lightpaths between node pair $(0, 9)$. Thus, the two wavelengths on link $\langle 6, 9\rangle$ can only be used by backup paths, and the two primary paths between node pair $(0, 9)$ need to traverse both links $\langle 4, 5\rangle$ and $\langle 5, 9\rangle$. Under share-path protection, the two backup paths between node pair $(0, 9)$ cannot share any wavelength along the entire paths because their corresponding primary paths between node pair $(0, 9)$ both traverse links $\langle 4, 5\rangle$ and $\langle 5, 9\rangle$). Under sub-path protection, node 4 segments each primary path between node pair $(0, 9)$ into two sub-paths (one in each domain). The fact that the two primary sub-paths in the right domain are not mutually diverse does not affect the two primary sub-paths in the left domain. As a result, the backup sub-paths for the two primary sub-paths in the left domain share the wavelengths along the entire paths because the two primary sub-paths in the left domain are fiber-disjoint. Hence, the optimal solution for the RWA under shared-path-protection constraints requires two more wavelength-links (along path $\langle 0, 3, 4\rangle$) than the optimal solution for the RWA under sub-path-protection constraints.

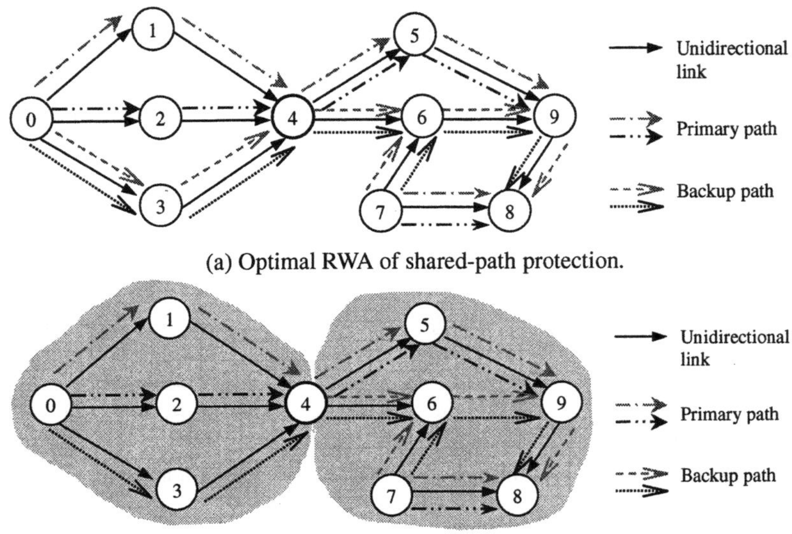

(a) Optimal RWA of shared-path protection.

(b) Optimal RWA of sub-path protection. Each cloud denotes a domain.

Figure 3.7. Sub-path protection outperforms shared-path protection in terms of resource utilization (note the difference along path $\langle 0, 3, 4 \rangle$.) Node 4 is wavelength convertible.

The rationale behind this property is that each sub-path is independent of the other sub-paths of the same lightpath. (Please refer to the example in Section 3.2.1.) This gain is achieved by the domain constraints, namely, the primary and backup paths of an inter-domain lightpath exit a domain (and enter another domain) through a common DBN and an intra-domain lightpath cannot use resource of other domains. As the size of the network or the number of lightpath requests increases, the correlation between backup paths (two backup paths cannot share wavelength if their corresponding primary paths are not mutually diverse) can become noticeable or even significant. Sub-path protection reduces this correlation by segmenting a long lightpath into several shorter ones, thus it increases the potential sharing between backup paths. Then, it would be expected that the difference in the number of wavelength-links per lightpath between sub-path protection and shared-path protection decreases as the number of lightpaths increases. The reason Fig. 3.6(b) does not show this trend is that we assume a non-blocking model, i.e., all the lightpath requests can be routed (the number of wavelengths on each link is large). The situation in Fig. 3.7 will not be the case if we have one more wavelength on each link. Thus, this increased backup sharing is more likely to happen when the network load is high.

3.6 Conclusion

We proposed and investigated sub-path protection for survivable lightpath provisioning and fast protection switching in optical WDM mesh networks. The main ideas of sub-path protection are to partition a large optical network into smaller domains, and to apply shared-path protection to the optical network while guaranteeing the autonomy of each domain. We proved that the RWA problem under shared-path protection for a given set of lightpath requests is NP-complete (even if the number of wavelength on each link is one), so is the problem of RWA under sub-path protection for a given set of lightpath requests since the former is a special case of the latter. Furthermore, we proved that the problem of finding optimal backup paths, under shared-path-protection constraints (or sub-path-protection constraints) for a set of lightpath requests whose primary paths are given, is still NP-complete. We mathematically formulated the RWA problem with sub-path-protection constraints for a given set of lightpath requests and developed a heuristic to find efficient solutions. While sub-path protection reduces the ability to find globally optimal solution due to domain partitioning, it increases backup sharing. The comparisons between sub-path protection and shared-path protection on a nationwide network with dozens of wavelengths per fiber show that, for a modest sacrifice in resource utilization, sub-path protection achieves improved survivability, much higher scalability, and significantly reduced fault-recovery time.

Please note that the performance of our sub-path protection largely depends on how a network is partitioned. The significant benefits–shorter recovery time and higher scalability–of sub-path protection come from the fact that the network is well partitioned. In our example, we chose the network partitions based on geography. In general, how to partition a network may depend on the topology and involve administrative issues. As a result, how to properly partition a network is expected to be a challenging problem and is left open for future research.

This chapter investigated from the network perspective sub-path protection to reduce the recovery time of shared-path protection. The next chapter explores from the connection perspective segment protection to reduce the recovery time of shared-path protection.

APPENDIX 3.A: NP-Completeness of RWA for Shared-Path Protection

Proof of Theorem 3.1: We reduce the known NP-complete problem, multi-source multi-destination edge-disjoint paths problem [Frieze, 2001] (MSMDED), to RWASPP. The decision version of MSMDED is stated as follows.

Instance: A graph $G = (V, E)$, and a set of D node pairs $(s_1, d_1), \cdots, (s_D, d_D)$ (D is an input).

Question: Is there a set of D mutually link-disjoint paths, one for each node pair?

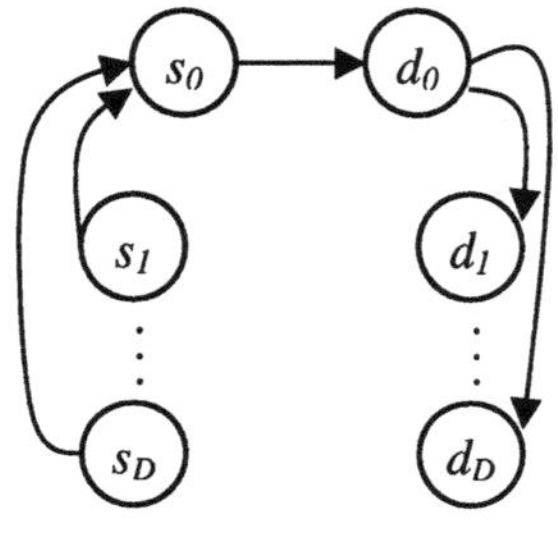

Figure 3.A.1. Construction of NP-complete proof.

It is easy to see that RWASPP $\in$ NP since a non-deterministic algorithm can guess a link-disjoint path pair for each node pair and check in polynomial time if the D link-disjoint path pairs found satisfy constraints C.1-C.4 (in Section 3.2.4).

We reduce MSMDED to RWASPP. Let $\tilde{G} = (\tilde{V}, \tilde{E}), \tilde{\mathcal{D}} = \{(s_k, d_k) | \ 1 \leq k \leq D, s_k, d_k \in \tilde{V}\}$ be an arbitrary instance of MSMDED. Construct an instance of RWASPP as follows: $W = 1, \mathcal{D} = \tilde{\mathcal{D}}$, and $G = (V, E)$, where $V = \tilde{V} \cup \{s_0, d_0\}(s_0, d_0 \notin \tilde{V}), E = \tilde{E} \cup \{\langle s_k, s_0 \rangle | \ 1 \leq k \leq D\} \cup \{\langle s_0, d_0 \rangle\} \cup \{\langle d_0, d_k \rangle | \ 1 \leq k \leq D\}$ (please refer to Fig. 3.A.1). Clearly, the construction can be accomplished in polynomial time.

Suppose there exist D mutually link-disjoint paths—$\tilde{p_1}, \cdots, \tilde{p}_D$, one for each node pair—in the instance of MSMDED. We construct the set of D link-disjoint path pairs, (p_p^k, p_b^k), as follows (please note that there is only one wavelength):

$$p_p^k = \tilde{p_k}, \quad p_b^k = \langle s_k, s_0, d_0, d_k \rangle \qquad \forall \, 1 \leq k \leq D$$

Then, Constraints C.1, C.2, and C.4 are satisfied since the set of paths $\{p_p^k \mid 1 \leq k \leq D\}$ is mutually link-disjoint; and Constraint C.3 is satisfied because backup paths use links only in E, not in $\tilde{E}$. Thus, the set of path pairs so constructed is a feasible solution to RWASPP.

Suppose the instance of RWASPP has a set of D link-disjoint path pairs, $\{(p_w^k, p_b^k) | \ 1 \leq k \leq D\}$, which satisfies Constraints C.1-C.4. Consider link $\langle s_0, d_0 \rangle$. It can be in only one of the following two states:

1 *Link $\langle s_0, d_0 \rangle$ is used by some primary path p_p^j.* Then, only p_p^j uses $\langle s_0, d_0 \rangle$ since $W = 1$. Construct the set of D paths in $\tilde{G}$ as follows.

$$\tilde{p_k} = \begin{cases} p_p^k & \text{if } k \neq j \\ p_b^k & \text{if } k = j \end{cases} \qquad \forall \, 1 \leq k \leq D$$

Clearly, there is one path for each node pair; by Constraint C.2, $\tilde{p_1}, \cdots, \tilde{p}_{j-1}, \tilde{p}_{j+1}, \cdots, \tilde{p}_D$, are mutually link-disjoint since $W = 1$; and by Constraint C.3, $\tilde{p_1}, \cdots, \tilde{p}_{j-1}, \tilde{p}_j, \tilde{p}_{j+1}, \cdots, \tilde{p}_D$ are mutually link-disjoint. Thus, the set of paths $\{\tilde{p_k} | \ 1 \leq k \leq D\}$ so found is a feasible solution to MSMDED.

2 *Link $\langle s_0, d_0 \rangle$ is not used by any primary path.* Let $\tilde{p_k} = p_p^k$. Then, there is one path for each node pair. By Constraint C.2, the set of paths $\{\tilde{p_k} \mid 1 \leq k \leq D\}$ is mutually link-disjoint because $W = 1$. Thus, the set of paths $\{p_k | \ 1 \leq k \leq D\}$ so found is a feasible solution to MSMDED.

This concludes our proof that RWA under shared-path-protection constraints is NP-complete. Thus, RWA under sub-path protection is also NP-complete since the former is a special case of the latter when the number of partitions, $|\mathcal{P}|$, is one.

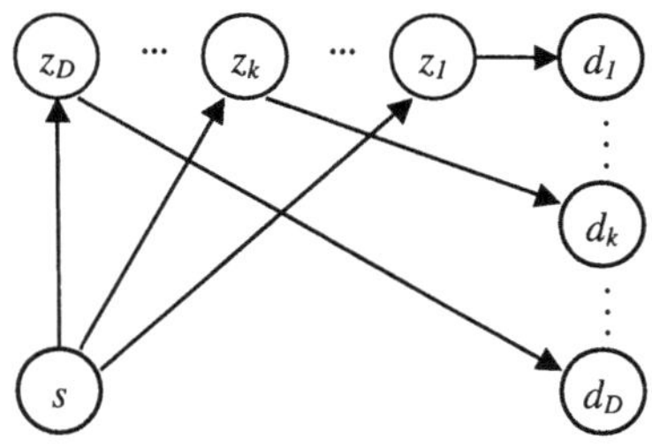

Figure 3.B.1. Reduce DSMT to OBR.

APPENDIX 3.B: NP-Completeness of Optimal Backup Routing (OBR)

Proof of Theorem 3.2: We reduce the directed Steiner minimal tree (DSMT) problem, which is NP-complete [Ramanathan, 1996], to OBR. The decision version of the DSMT problem is defined as follows.

DSMT: Given a weighted, directed graph $G = (V, E, C)$, a set of nodes $\tilde{V} \subset V$, an integer c, and a source node $s \notin \tilde{V}$, does there exist a directed Steiner tree (DST) $\mathcal{T}$ such that there exists a directed path in $\mathcal{T}$ from s to every vertex in $\tilde{V}$, and the cost of $\mathcal{T}$, $C(\mathcal{T})$, is no more than c?

OBR $\in$ NP since a non-deterministic algorithm can guess a backup path for each lightpath and check in polynomial time if the primary and backup paths satisfy Constraints C.1-C.4 (in Section 3.2.4) and the cost of all the backup paths is no more than c.

Given an arbitrary instance of DSMT $G = (V, E, C)$, $\tilde{V}$, and c, construct an instance of OBR $G' = (V', E', C')$, $W = 1$, $\mathcal{D}$, and p_p^k as follows (let $D = |\tilde{V}|$):

$$\begin{aligned}
V' &= V \cup \{z_1, ..., z_D\} \\
E' &= E \cup \{sz_k \mid \forall\, 1 \le k \le D\} \cup \{z_k d_k \mid \forall\, d_k \in \tilde{V}\} \\
C'(ij) &= \begin{cases} C(ij) & \text{if } ij \in E \\ 0 & \text{otherwise} \end{cases} \\
\mathcal{D} &= \{(s, d_k) \mid \forall\, d_k \in \tilde{V}\} \\
p_p^k &= \langle s, z_k, d_k \rangle \quad \forall\, d_k \in \tilde{V}
\end{aligned}$$

This construction is shown in Fig. 3.B.1. Clearly, the transformation can be done in polynomial time.

Suppose the instance of DSMT can find a DST $\mathcal{T}$ of cost no more than c. Let the backup path of lightpath k, p_b^k, be the path from s to d_k, using links only in $\mathcal{T}$. First, Constraints C.1-C.4 are satisfied since primary paths only use newly added links and none of the two primary paths traverse any common link; thus, the backup paths can share any wavelength along any link in E. Second, the set of backup paths so found has cost no more than c. Thus, the set of backup paths so found is a feasible solution for OBR.

Suppose the instance of OBR can find a set of backup paths, $\{p_b^k \mid \forall\, 1 \le k \le D\}$, such that it has cost no more than c. If $\{p_b^k\}$ forms a tree, then the tree so formed is a feasible solution for DSMT. Otherwise, we can repeatedly remove a link (which is of nonnegative cost) from the set of links used by $\{p_b^k\}$ until it forms a tree rooted at s, then the tree so formed is a feasible solution to DSMT, and it has cost no more than c.

This concludes our proof that OBR is NP-complete.

Chapter 4

SEGMENT PROTECTION FOR BANDWIDTH EFFICIENCY AND DIFFERENTIATED QUALITY OF PROTECTION

4.1 Introduction

We consider the problem of dynamic survivable lightpath provisioning against single node (crossconnect) and single link (fiber) failures. Specifically, we focus on shared protection (because of its desirable resource efficiency) with the assumptions that existing lightpaths cannot be disturbed and no knowledge of future arrivals is available at the time of provisioning the current lightpath request. While we consider full wavelength-convertible networks here, the extension to the wavelength-continuous case is straightforward.

One possible limitation of the widely studied shared-path protection is that backup paths may sometimes become longer due to backup sharing [Bouillet et al., 2002b]. Consequently, protection-switching time may increase because of longer backup paths. The relation between backup sharing and backup-path hop distance for path protection have been shown to be that one trades off another in [Bouillet et al., 2002b, Xiong et al., 2003].

Furthermore, lightpath requests may have differentiated protection-switching-time requirements. For example, lightpaths carrying voice traffic may require 50-ms protection-switching time while lightpaths carrying data traffic may have a wide range of protection-switching-time requirements. Due to the path-wise node-/link- disjoint nature of path protection, shared-path protection may not provision lightpath requests according to their protection-switching-time requirements effectively in practical-sized networks [Li et al., 2003, Ou et al., 2002]. Clearly, proper mechanisms are needed to provision such lightpath requests in a resource-efficient manner.

Motivated by the above considerations, we first unify various forms of segment protection into generalized segment protection (GSP) and propose an effective heuristic in Section 4.2. Then, based on GSP, we present a new

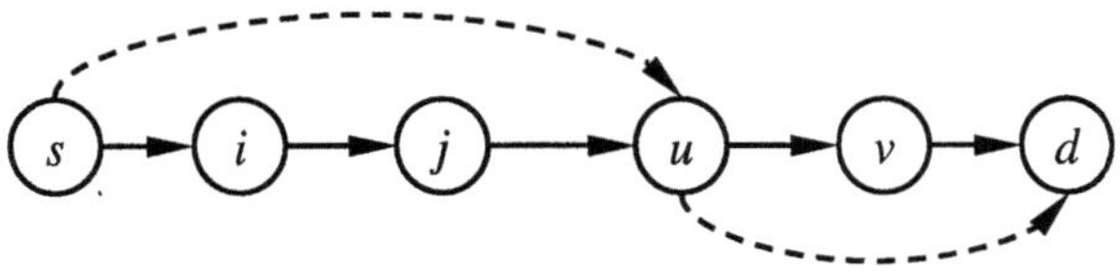

(a) Non-overlapping segment protection as in [Anand et al., 2002, Sridharan et al., 2002, Todimala and Ramamurthy, 2002].

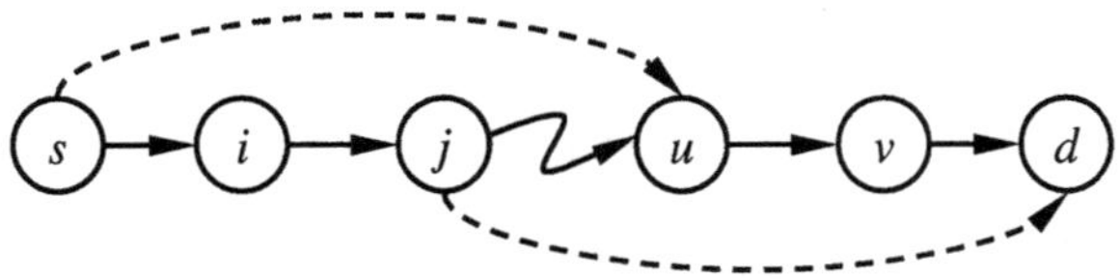

(b) Overlapping segment protection as in [Gummadi et al., 2003, Ho and Mouftah, 2002].

Figure 4.1. Various forms of segment protection. The solid lines from node s to node d represent the working path, and the dashed lines represent the backup segments. While only two segments are shown in these illustrations, in general, a path may employ many segments. Also, each backup segment may have several additional intermediate nodes, which are not shown here to avoid cluttering.

and effective approach to provisioning lightpath requests according to their protection-switching-time requirements while taking into account backup sharing in Section 4.3.

While our focus is on optical WDM network, in which the bandwidth requirement of a lightpath request is one wavelength, our approaches can also be directly applied to MPLS networks for provisioning restorable, bandwidth-guaranteed connections of differentiated bandwidth granularities with appropriate adjustments.

4.2 Generalized Segment Protection

Below, we unify various forms of segment protection into generalized segment protection (GSP) in Section 4.2.1, design an efficient heuristic in Section 4.2.2, and demonstrate the effectiveness of GSP in Section 4.2.3.

4.2.1 Generalized Segment Protection

Various forms of segment protection were reported in [Anand et al., 2002, Gummadi et al., 2003, Ho and Mouftah, 2002, Sridharan et al., 2002, Todimala and Ramamurthy, 2002]. The approaches proposed in [Anand et al., 2002, Sridharan et al., 2002, Todimala and Ramamurthy, 2002] addressed single-link failures by dividing a working path into a sequence of non-overlapping segments and protecting each such segment individually. As shown in Fig. 4.1(a), the lightpath from node s to node d is partitioned into two non-overlapping segments: one with working segment $\langle s, i, j, u \rangle$ and backup segment $\langle s, u \rangle$; another with working segment $\langle u, v, d \rangle$ and backup segment $\langle u, d \rangle$. (Note that each backup segment may have several additional intermediate nodes, which are not shown in Fig. 4.1 to avoid cluttering.) When a failure occurs, only the affected segment performs protection switching and the other unaffected segments are oblivious to the failure. For example, in Fig. 4.1(a), if link $\langle i, j \rangle$ fails, the working segment $\langle s, i, j, u \rangle$ switches to its backup segment $\langle s, u \rangle$, and the other working segment $\langle u, v, d \rangle$ is unaware of the failure. Node failures are not accommodated in these approaches as consecutive non-overlapping segments share the same node failure, e.g., node u in Fig. 4.1(a).

Even though node failures do not occur as often as link failures, they need to be carefully treated because the impact of node failures is much more disastrous than that of link-failures. The work in [Gummadi et al., 2003, Ho and Mouftah, 2002] handled single-node/link failures by dividing a working path into a sequence of overlapping segments and protecting each such segment separately. As shown in Fig. 4.1(b), the lightpath from node s to node d is partitioned into two overlapping segments: one with working segment $\langle s, i, j, \cdots, u \rangle$ and backup segment $\langle s, u \rangle$; and another with working segment $\langle j, \cdots, u, v, d \rangle$ and backup segment $\langle j, d \rangle$.

We unify the above approaches into generalized segment protection, which is almost the same as the overlapping segment protection shown in Fig. 4.1(b) except that node j and node u can be the same node. GSP differs from the previous approaches in that it can dynamically divide a working path into multiple segments while accommodating backup sharing, as will be elaborated later in Section 4.2.2.

Similar to segment protection, GSP has a number of advantages compared to path protection.

The end-to-end protection entity is *a segment* in segment protection as opposed to *a path* in path protection. When a failure occurs along a working path (segment), the source node of that path (segment) switches to its backup. Since a segment is typically shorter than a path in terms of hop count, segment protection is expected to have shorter protection-switching time.

Meanwhile, two segments (or two lightpaths in path protection) can share backup wavelength links as long as their working segments (or working paths in path protection) do not share the same node/link failure. Since, in general, a segment is shorter than a path, the probability of two working segments sharing the same risk is typically lower than the probability of two working paths sharing the same risk. As a result, segment protection can have better backup sharing compared to shared-path protection.

Furthermore, segment protection has more flexibility in routing compared to path protection since path protection is a special case of segment protection in which every lightpath has exactly one segment. Apart from that, it is clear that the longer the working path is, the more difficult it is to find a node-disjoint backup path [Liu and Tipper, 2001]. Later, we shall show that an improperly-selected working path can partition a network and no end-to-end (with respect to lightpath) node-disjoint backup path can be found. Thus, it is desirable to have shorter working path to achieve routing flexibility, as is the case in segment protection.

Next, we design an effective heuristic to compute a route for an incoming connection request.

4.2.2 The GSP Heuristic

Upon the arrival of a new lightpath request, the network management system needs to compute a working path l_w and a list of backup segments $\{l_b^i\}$, which divide the working path into overlapping segments $\{l_w^i\}$ such that l_w^i and l_b^i are node-/link- disjoint. New backup segments $\{l_b^i\}$ can share wavelength links with existing backup segments as well as among themselves. Unfortunately, it is NP-hard to determine if there exists an eligible solution as we have proved the NP-completeness of the existence version of shared-path protection, which is a special case of segment protection with the number of segments being one [Ou et al., 2004]. As a result, we resort to a heuristic.

Below, we first define the notations, and then present a practical heuristic which, upon the arrival of a new lightpath request, dynamically divides a judiciously-selected working path into multiple overlapped working segments and computes a backup segment for each working segment while accommodating backup sharing.

4.2.2.1 Notations

A network is represented as a weighted, directed graph $G = (V, E, C, \lambda)$, where V is the set of nodes, E is the set of unidirectional fibers (referred to as links), $C : E \rightarrow R^+$ is the cost function for each link (where R^+ denotes the set of positive real numbers), and $\lambda : E \rightarrow Z^+$ specifies the number of wavelengths on each link (where Z^+ denotes the set of positive integers).

A conflict set is associated with a link to identify the sharing potential between backup segments[1]. The conflict set ν_e for link e defines the set of nodes traversed by such working segments whose backup segments utilize wavelengths on link e. The conflict set ν_e for link e can be represented as an integer set, $\{\nu_e^u | \ \forall u \in V, 0 \leq \nu_e^u \leq \lambda(e)\}$, where ν_e^u specifies the number of working segments which traverse node u and are protected by link e (or, in other words, their

[1] The conflict set is similar to the conflict vector in [Mohan et al., 2001], the aggregated square matrix in [Liu et al., 2001], and the "bucket" link metric in [Su and Su, 2001b].

corresponding backup segments traverse link e). The number of wavelengths reserved for backup segments on link e is thus $\nu_e^* = \max_{\forall u}\{\nu_e^u\}$. Clearly, the union of the conflict sets for all the links aggregates the per-segment-based information, and the size of the conflict set depends only on the number of links, not on the number of segments.

4.2.2.2 GSP Heuristic

The route-computation approaches in [Anand et al., 2002, Ho and Mouftah, 2002, Sridharan et al., 2002, Todimala and Ramamurthy, 2002] partition a working path in a fixed manner, e.g., every working path is divided into a constant number of segments or into multiple segments of equal hop count. A flexible partitioning approach in [Gummadi et al., 2003] dynamically divides a working path into overlapping segments, but does not take into account backup sharing. Our GSP heuristic extends the idea in [Gummadi et al., 2003] to incorporate backup sharing and to facilitate partitioning a working path into overlapping working segments.

Our GSP heuristic is specified in detail in Algorithm 4.1. In Algorithm 4.1, K is an input constant representing the maximum number of candidate working paths; and ϵ is a small constant such as 0.01. The value of ϵ is used to control the degree of backup sharing: smaller values encourage backup sharing and larger values discourage backup sharing [Bouillet et al., 2002b].

The basic ideas of our GSP heuristic are as follows.

1) Select a candidate working path l_w^k and transform the original graph based on l_w^k in a way such that any path link-disjoint to l_w^k in the transformed graph can be mapped back to the original graph and decomposed into a set of backup segments $\{l_b^{k,i}\}$, which partitions l_w^k into multiple working segments $\{l_w^{k,i}\}$ where consecutive segments overlap by *at least* one hop and $l_w^{k,i}$ is node-/link-disjoint to $l_b^{k,i}$.

2) Consider the worst-case-scenario backup sharing when computing the backup segments but precisely allocate backup resources after the backup segments are computed. Basically, when computing the backup segments for working path l_w^k, we consider the worst case (as far as backup sharing is concerned) where l_w^k is one segment since the working segments cannot be determined without the backup segments. Later, when the list of backup segments $\{l_b^{k,i}\}$ is computed, backup sharing is performed on a per-segment basis in Steps 4 and 5 as the list of working segments $\{l_w^{k,i}\}$ can be determined.

Figure 4.2 highlights some distinct features of our GSP heuristic. When a new lightpath request from node s to node d arrives at the network state shown in Fig. 4.2(a), the only candidate working path is $\langle s,i,j,u,v,d\rangle$. We observe that shared-path protection cannot find a solution as there is no path which is end-to-end disjoint to path $\langle s,i,j,u,v,d\rangle$.

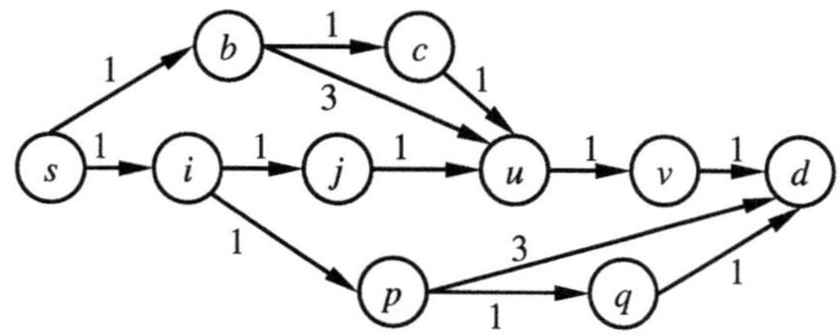

(a) A network state $G = (V, E, C, \lambda)$ in which two lightpaths—lightpath one with working path $\langle b, c, u \rangle$ and backup path $\langle b, u \rangle$ and lightpath two with working path $\langle p, q, d \rangle$ and backup path $\langle p, d \rangle$—are already set up. Every link has one wavelength.

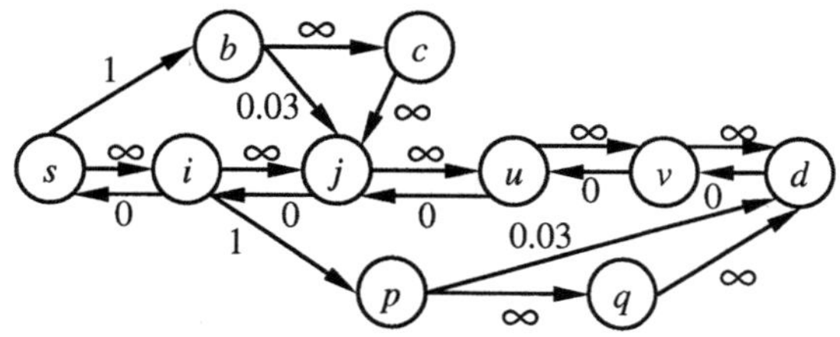

(b) $G' = (V, E', C', \lambda)$ for working path $\langle s, i, j, u, v, d \rangle$ (assuming $\epsilon = 0.01$ in Algorithm 4.1).

Figure 4.2. Illustration of the GSP heuristic. The number besides a link represents the cost of that link.

However, our heuristic can find a solution. Figure 4.2(b) shows the transformed graph $G' = (V, E', C', \lambda)$. Following Step 2b in Algorithm 4.1, the minimal-cost path $\langle s, b, j, i, p, d \rangle$ will be computed as l_b^k, which will then be mapped back to G and decomposed as two backup segments $\langle s, b, u \rangle$ and $\langle i, p, d \rangle$ (note that link $\langle b, j \rangle$ in G' was constructed from link $\langle b, u \rangle$ in G). This example highlights that our GSP heuristic can dynamically divide a working path into multiple arbitrary overlapping working segments and protect each working segment separately while accommodating backup sharing.

4.2.2.3 Computational Complexity

Algorithm 4.1 has a computational complexity of $O(K \cdot (|V|^3 + |E|))$. In particular, the complexity of Step 1 is $K \cdot |V|^3$; the complexity of Step 2 is $O(K \cdot |E|)$ (the computational complexities of Steps 2a, 2b, 2c, and 2d are $O(|E|), O(|V|^2), O(|E|)$, and $O(|E|)$, respectively); the complexities of Steps 3, 4, 5, and 6 are $O(1), O(|E|), O(|E|)$, and $O(1)$. If $K = 1$, then the complexity of Step 1 can be reduced from $K \cdot |V|^3$ to $O(|V|^2)$. Consequently, the complexity of Algorithm 4.1 can be reduced to $O(|V|^2 + |E|)$, which is the complexity of shortest-path algorithms.

Algorithm 4.1 GSP

Input: $G = (V, E, C, \lambda)$, $\nu = \{\nu_e \mid e \in E\}, s, d, K$

Output: A working path l_w and a list of backup segments $\{l_b^i\}$ which partitions l_w into overlapping segments $\{l_w^i\}$ such that l_w^i & l_b^i are node-/link- disjoint; otherwise NULL if no eligible solution is found.

1 **select candidate working paths**: compute up to K minimal-cost paths $L_w = \{l_w^k \mid 1 \leq k \leq K\}$ in G from s to d based on Yen's K-shortest paths algorithm [Yen, 1971] subject to the constraint that every hop along a path should have at least one free wavelength; return NULL if L_w is empty

2 **compute backup segments for each candidate working path** l_w^k in L_w as follows:

(a) transform $G = (V, E, C, \lambda)$ to $G' = (V, E', C', \lambda)$:

i define link-cost function $C''(e)$ for $e \in E$:

$$C''(e) := \begin{cases} \infty & \text{if } l_w^k \text{ traverses link } e\text{, or } \nu_e^* = \nu_e^u \text{ for some node } u \\ & \text{along } l_w^k \text{ and link } e \text{ does not have any free wavelength} \\ \epsilon \cdot C(e) & \text{if for any node } u\ (u \neq s, d) \text{ traversed by } l_w^k,\ \nu_e^u < \nu_e^* \\ C(e) & \text{otherwise} \end{cases}$$

ii define link-set E' and $C'(e)$ for $e \in E'$:

- $\forall \langle u, v\rangle \in E \wedge u \neq s \wedge u \neq d$, if l_w^k traverses node v but not node u, then add link $\langle u, p\rangle$ to E', where node p is v's immediate predecessor along l_w^k, and let $C'(\langle u, p\rangle) = C''(\langle u, v\rangle)$; otherwise, add $\langle u, v\rangle$ to E' and let $C'(\langle u, v\rangle) = C''(\langle u, v\rangle)$
- $\forall \langle u, v\rangle \in E$ and l_w^k traverses $\langle u, v\rangle$, if $\langle v, u\rangle \notin E$, then add $\langle v, u\rangle$ into E' and let $C'(\langle v, u\rangle) = 0$

(b) compute a least-cost path l_b^k from s to d in G'

(c) map l_b^k back to G and decompose l_b^k into a list of backup segments $\{l_b^{k,i}\}$ which partitions l_w^k into overlapped working segments $\{l_w^{k,i}\}$

(d) $\forall i$, compute the amount of fresh wavelength links backup segment $l_w^{k,i}$ consumes: for any link e that $l_b^{k,i}$ traverses, if $\nu_e^u = \nu_e^*$ for some node u along $l_w^{k,i}$ (excluding the source and destination nodes of $l_w^{k,i}$), then increase the amount of fresh wavelength links by one

3 select the pair $\langle l_w^k, \{l_b^{k,i}\}\rangle$ of minimal cost; return NULL if no such pair exists

4 **allocate resources for** $\langle l_w^k, \{l_b^{k,i}\}\rangle$: allocate a new wavelength along l_w^k and update backup wavelengths for every $l_b^{k,i}$: for any link e that $l_b^{k,i}$ traverses, if $\nu_e^u = \nu_e^*$ for some node u along $l_w^{k,i}$ (excluding the source and the destination nodes of $l_w^{k,i}$), then reserve one more wavelength on link e as backup resources

5 **update the conflict set**: $\forall i$, update conflict set associated to links traversed by $l_b^{k,i}$: for every link e that $l_b^{k,i}$ traverses, $\nu_e^u \leftarrow \nu_e^u + 1$ for every node u along $l_w^{k,i}$ (excluding the source node and the destination node of $l_w^{k,i}$)

6 return l_w^k and $\{l_b^{k,i}\}$

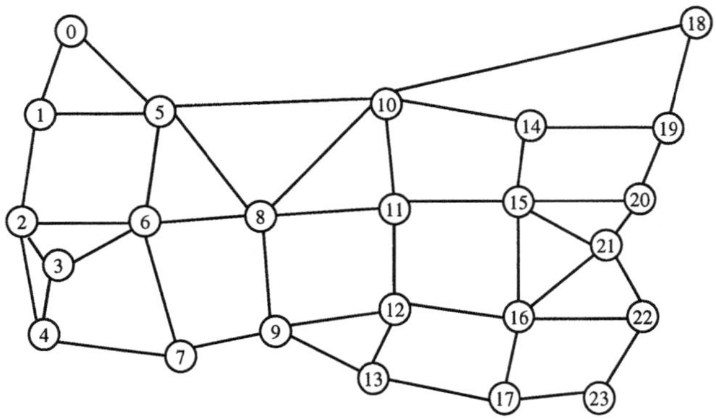

Figure 4.3. A representative topology whose average hop distance is about 2.99 and average nodal degree is about 3.58.

4.2.3 Illustrative Numerical Results

We now quantitatively evaluate GSP. We simulate a dynamic network environment with the assumptions that the lightpath-arrival process is Poisson and the lightpath-holding time follows a negative exponential distribution. In every experiment, 10^6 lightpath requests are simulated; they are uniformly distributed among all node pairs; average lightpath-holding time is normalized to unity; the cost of any link is unity; and our example network topology with 16 wavelengths per fiber is shown in Fig. 4.3. For the results shown in this section, $\epsilon = 0.01$ since we aim to maximize backup sharing. We remark that more results from different topologies also led to the same observations. Those results are not shown here.

Below, we compare GSP to shared-path protection, which is widely considered to be the most resource-efficient protection scheme so far. Since shared-path protection is a special case of GSP, the heuristic for shared-path protection differs from Algorithm 4.1 only in Step 2a and Step 2b (some steps such as Step 2c can be removed for efficiency because the number of segments in shared-path protection case is always one). Step 2a is modified as follows: temporarily remove all the nodes traversed by l_w^k (except node s and node d) and all the links sourced/sunk at the removed nodes. Step 2b is modified as follows: compute a minimal-cost path l_b^k from node s to node d in G with link-cost function C''.

4.2.3.1 Blocking Probability

Figure 4.4 compares the blocking probability of GSP to that of shared-path protection for $K = 1, 2$, and 3. We make the following observations: (a) GSP has much lower blocking probability than shared-path protection for the same K. This is because GSP can achieve better backup sharing and have more

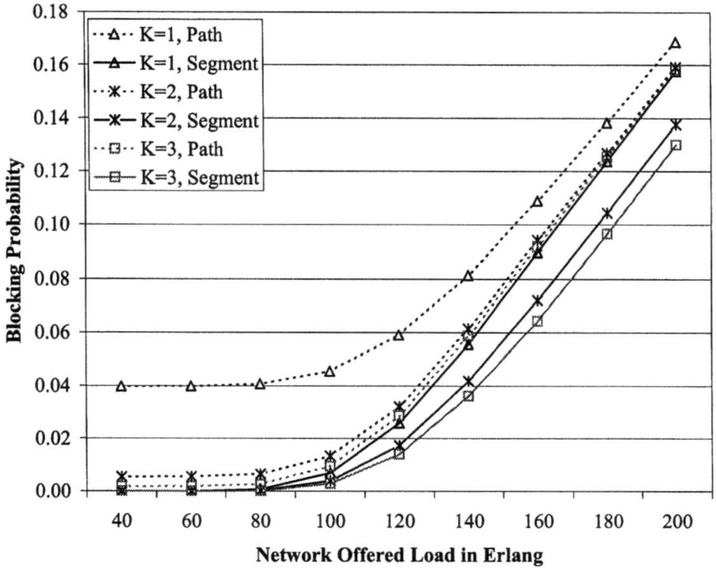

Figure 4.4. Blocking probability. The average link utilization for 40 Erlangs is about 17% and for 200 Erlangs is about 65%.

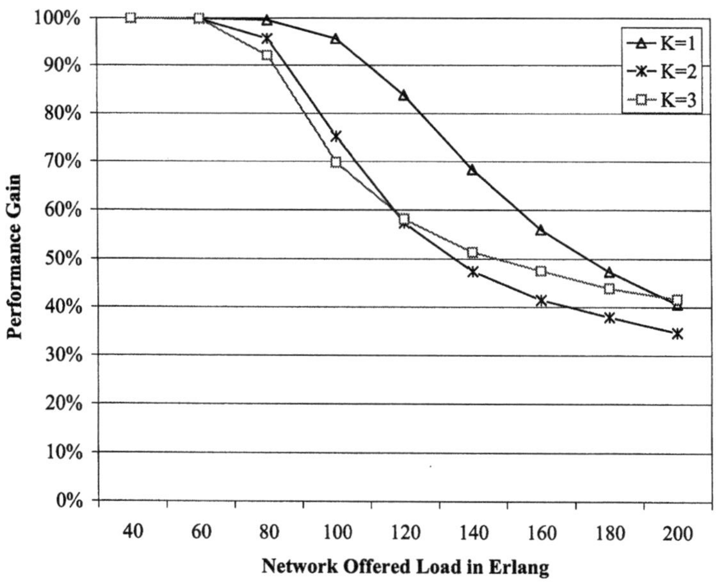

Figure 4.5. Performance gain of GSP over shared-path protection.

flexibility in routing, as discussed earlier in Section 4.2.1. (b) When K increases from one to two, the reduction in blocking probability for both GSP and shared-path protection is significant while the reduction is only marginal when K further increases from two to three. This is basically the effect of alternate routing: the performance improvement is significant when the number of alternate routes increases from one to two, and the improvement is marginal or negligible when the number of alternate routes increases further [Ramamurthy and Mukherjee, 2002]. (c) Shared-path protection can have modest blocking probability even at low load, e.g., 40 Erlangs, when $K = 1$. Similar effect was also observed in [Liu and Tipper, 2001, Ou et al., 2004]. The reason is that an improperly-selected working path can disconnect the network and a backup path, which should be end-to-end node-/link- disjoint to the working path, can not be found. In some cases, the least-cost paths turn out to be improper working paths. For example, the least-cost path from node 0 to node 13 in Fig. 4.3 is $\langle 0, 5, 8, 9, 13 \rangle$. Clearly, there is no path node-disjoint to $\langle 0, 5, 8, 9, 13 \rangle$ from node 0 to node 13. As a result, a lightpath request from node 0 to node 13 may be blocked in shared-path protection when $K = 1$. However, in GSP, backup segments $\langle 0, 1, 2, 6, 8 \rangle$ and $\langle 5, 10, 11, 12, 13 \rangle$ form a valid solution among others.

4.2.3.2 Performance Gain

Performance gain is defined as the percentage of lightpath requests which are blocked in shared-path protection but can be accepted by GSP. Performance gain can be calculated as follows. Whenever shared-path protection needs to block a lightpath request, we apply GSP to check whether the same lightpath request can be provisioned under the same network state (but we do not set up the lightpath request even if it can be provisioned). Figure 4.5 shows that GSP achieves significant performance gain over shared-path protection. This is because the routing constraint in GSP is node-/link- disjoint segment-wise; but, in shared-path protection, it is node-/link- disjoint path-wise, as discussed earlier in Section 4.2.1.

4.2.3.3 Protection-Switching Time

For shared-path protection, protection-switching time for a lightpath can be calculated based on the hop count of the working and backup paths of the lightpath, as shown in [Assi et al., 2003, Ou et al., 2002, Ramamurthy et al., 2003]. For GSP, protection-switching time for a segment can be calculated based on the hop count of the working and backup segments using the same methodology as in [Assi et al., 2003, Ou et al., 2002, Ramamurthy et al., 2003] since the protection entity is a segment. Figure 4.6 shows that the average hop count of working and backup segments in GSP is much smaller than the average hop count of working and backup paths in shared-path protection. As a result, protection-switching for GSP is faster than that of shared-path protection.

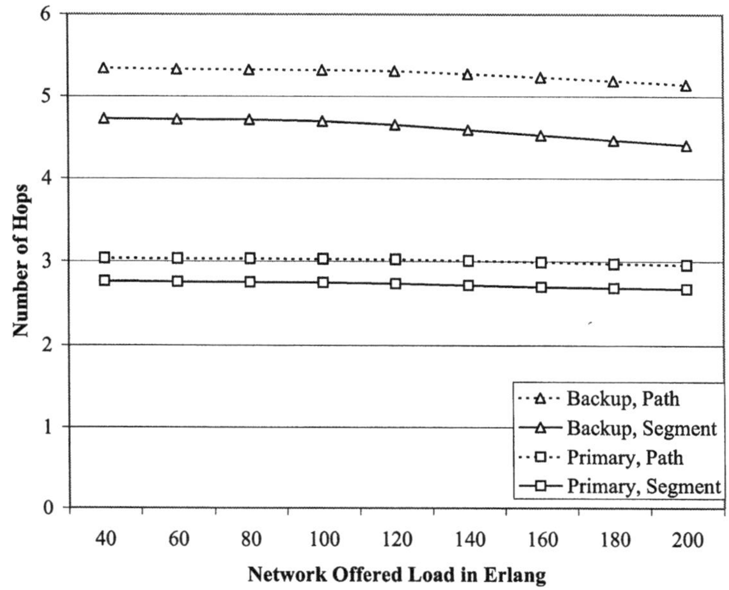

Figure 4.6. Working/backup hop count ($K = 2$). The plots for $K = 1$ and $K = 3$ are similar to the one for $K = 2$ here.

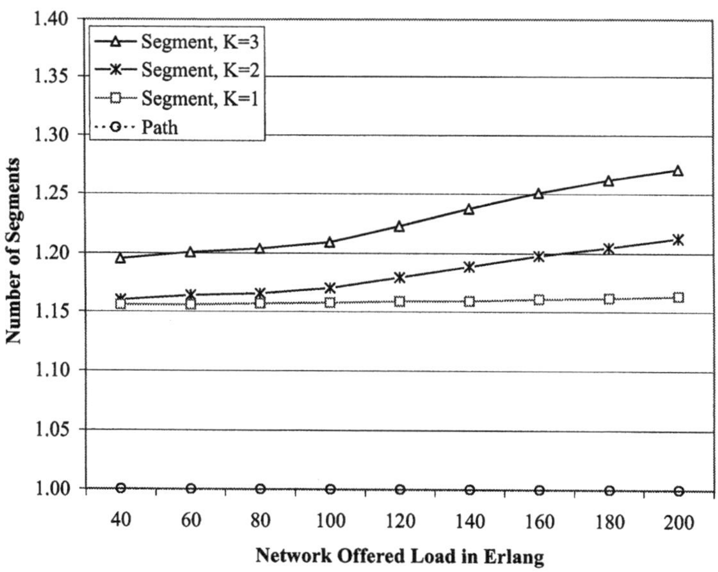

Figure 4.7. Number of segments per lightpath.

4.2.3.4 Control and Management Complexity

The control and management complexity might be higher in GSP than in shared-path protection since the number of segments is typically more than

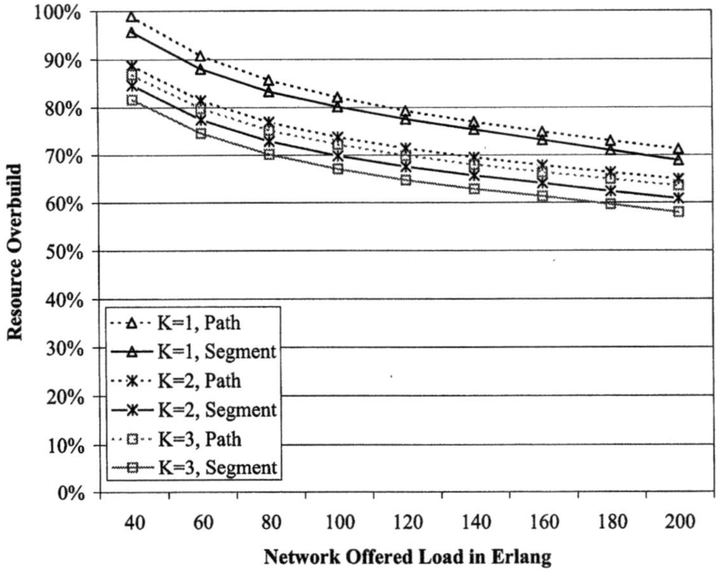

Figure 4.8. Resource overbuild.

the number of lightpaths. Figure 4.7 plots the average number of segments per lightpath, which is quite small (about 1.2) in this numerical example. As a result, the increase in control and management complexity is modest. Furthermore, by aggregating the per-segment information, we find that the size of the conflict set, which was defined in Section 4.2.2, depends only on the number of links, not on the number of segments.

4.2.3.5 Resource Efficiency

Resource overbuild, defined as the amount of wavelength links consumed by backup paths over the amount of wavelength links utilized by working paths [Li et al., 2002], indicates the amount of extra resources needed for providing protection as the percentage of the amount of resources required without protection. Typically, it is desirable to have lower resource overbuild because lower resource overbuild implies higher backup sharing. Figure 4.8 shows that GSP has lower resource overbuild than shared-path protection. The fact that a segment is shorter than a path contributes to increased backup sharing, and thus, decreased resource overbuild.

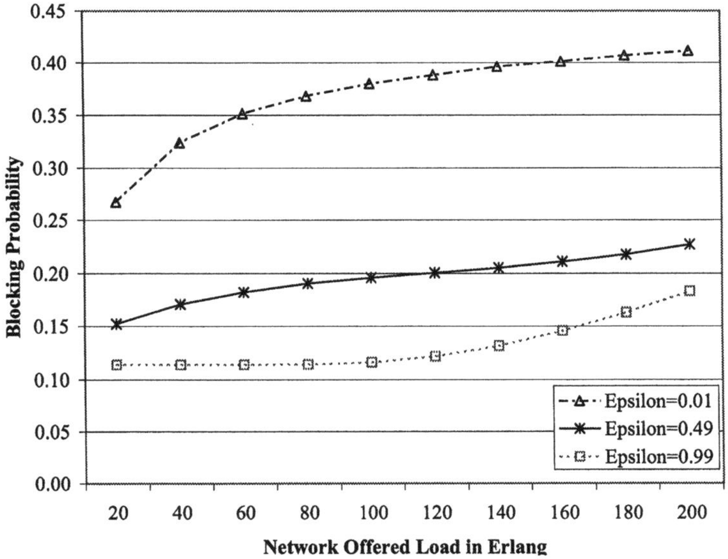

Figure 4.9. Shared-path protection ($K = 2$ and $H_b = 6$).

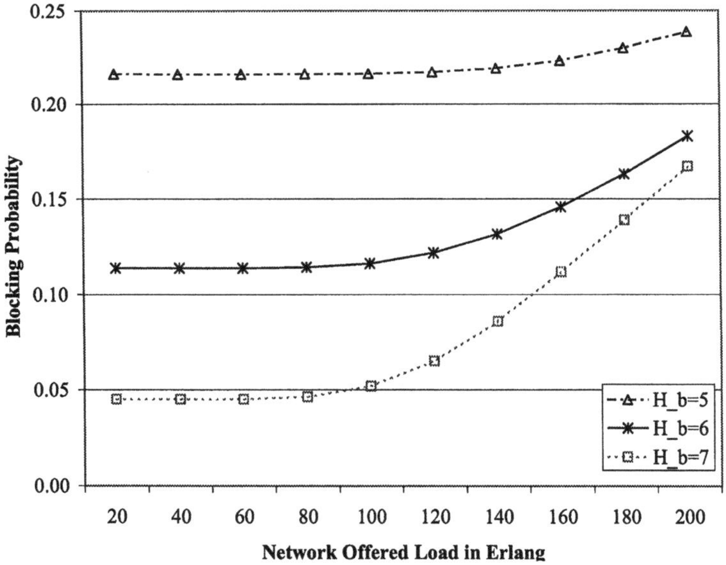

Figure 4.10. Shared-path protection ($K = 2$ and $\epsilon = 0.99$).

4.3 Providing Differentiated Quality of Protection (QoP) Based on Generalized Segment Protection

GSP can be employed for provisioning differentiated quality of protection (QoP). Here, we focus on one of the most important QoP parameters, namely protection-switching time. The protection-switching time of a shared-path protected lightpath can be calculated from the hop count of the working/backup paths, as shown in [Assi et al., 2003, Ou et al., 2002, Ramamurthy et al., 2003]. Therefore, we consider QoP in terms of hop count.

Below, we first argue why new mechanisms are needed to provision differentiated QoP in Section 4.3.1; in Section 4.3.2, we present a new approach to provision lightpath requests according to their QoP requirements; in Section 4.3.3, we evaluate the performance of our approach.

4.3.1 Motivation

Lightpath requests may have differentiated protection-switching-time requirements. For example, lightpaths carrying voice traffic may require 50-ms protection-switching time while lightpaths carrying data traffic may have a wide range of protection-switching-time requirements. While some lightpath requests (which carry mission-critical information) can be dedicated protected, it is not economically viable to provide dedicated protection to each lightpath request due to its excessive resource requirement.

Below, we show that shared-path protection may not be able to provide the desired level of protection-switching time either. Let us consider a simple case in which the backup-path hop count of any lightpath cannot exceed a constant H_b (ignoring the constraint on working path for now). A lightpath request will be blocked if the computed backup path is longer than H_b hops. Figure 4.9 plots the blocking probability of shared-path protection for $H_b = 6$ for the network shown in Fig. 4.3 with different values of ϵ. Recall that ϵ is the parameter used by the link-cost function in computing a shared backup path. Figure 4.9 confirms the conclusion in [Bouillet et al., 2002b, Xiong et al., 2003] that a larger value of ϵ leads to shorter backup path but decreased backup sharing, and a smaller value of ϵ leads to increased backup sharing but longer backup path (results for other values of ϵ follow the same trend, so they are not shown here). Please note that $H_b = 6$ for this network is reasonably large since the average backup-path hop count is about 5.2 for $\epsilon = 0.01$ as shown in Fig. 4.6. However, regardless of the values of ϵ, the blocking probability in Fig. 4.9 is quite high. The main reason is that some lightpath requests are blocked because their backup paths span more than $H_b = 6$ hops. Figure 4.10 shows the impact of H_b on shared-path protection. While the blocking probability drops significantly as H_b increases, the blocking for $H_b = 7$, which is quite large, is still unacceptable. (Please note that we chose $\epsilon = 0.99$ in Fig. 4.10 to discourage detouring of backup paths. The blocking probability of shared-path protection with hop-count constraint will be even worse if we chose a smaller value of ϵ.) As network size grows, it is clear that shared-path protection cannot achieve reasonable blocking for practical values of H_b due to its fundamental limitation: the backup path has to be end-to-end node-/link- disjoint to the working path.

Obviously, new mechanisms are needed. Below, we introduce more intelligence into GSP for provisioning lightpath requests to support differentiated QoP.

4.3.2 GSP_QoP Heuristic

We present a heuristic which applies GSP in a way such that the hop count of any backup segment is no more than H_b. For a candidate working path l_w^k, our heuristic, called GSP_QoP, performs the following recursive procedure to compute a list of eligible backup segments. 1) Starting from node s, compute a least-cost path to all the other nodes along l_w^k, where the cost function is C_1 defined in Algorithm 4.2. 2) Starting from node d and following the reverse direction of l_w^k, find the first node v which satisfies the constraint that the least-cost path from node s to node v is of at most H_b hops. 3) If node v is the destination node d, the heuristic succeeds and terminates; otherwise, starting from all the nodes between node s and node v (excluding nodes s and v) along l_w^k, recursively apply the above procedure (if there is no node between node s and node v, the heuristic fails).

Our GSP_QoP heuristic is specified in detail in Algorithm 4.2. For a node $u \in V$, $PC(u)$ denotes the cost of the least-cost path destined to node u; $HC(u)$ represents the hop count of the least-cost path; and $PH(u)$ records the previous hop along the least-cost path. For a path l_w, $Head(l_w)$ returns the first node along l_w.

We make the following remarks. 1) In Algorithm 4.2, the candidate working path l_w^k is given. This is just for the purpose of simplifying the presentation; in our implementation, we dynamically compute K candidate working paths as in Algorithm 4.1, execute Algorithm 4.2 for each candidate working path, and select the working path and the list of backup segments of minimal cost.

2) There are two objectives in computing a node-/link- disjoint backup segment. Objective one is to find a backup segment of hop count no more than H_b. Objective two is to select the backup segment of least cost. In general, constraint-based path selection with multiple objectives is NP-complete [Kuipers et al., 2002].

3) Backup sharing in this case is more tricky as the situation shown in Fig. 4.11 can arise. In the case of GSP (without the constraint of H_b), the path $\langle s, x, p, q, y, d\rangle$ could be a valid backup segment and this type of situation may not occur typically. In the presence of H_b, the path $\langle s, x, p, q, y, d\rangle$ is not valid when $H_b = 4$. However, the two segments $\langle s, x, p, q, u\rangle$ and $\langle j, p, q, y, d\rangle$ so formed are still valid. Our heuristic accommodates this type of backup sharing in Steps 3 and 7 since the freshly-reserved backup wavelengths for a newly-computed backup segment is used for computing later backup segments for the same lightpath request.

Algorithm 4.2 GSP_QoP

Input: $G = (V, E, C, \lambda)$, $\nu = \{\nu_e \mid e \in E\}$, s, d, a candidate working path l_w^k

Output: a list of backup segments $\{l_b^i\}$, each of which spans no more than H_b hops and they collectively partition l_w^k into overlapped segments $\{l_w^{k,i}\}$ such that $l_w^{k,i}$ and l_b^i are node-/link-disjoint; otherwise, NULL if no such list is found.

1 $S \leftarrow \{s\}, L_b \leftarrow \phi, l_w \leftarrow l_w^k, i \leftarrow 0$

2 $V' \leftarrow V$; $\forall u \in S, PC(u) \leftarrow 0, HC(u) \leftarrow 0, PH(u) \leftarrow NULL$; $\forall u \in V \wedge u \notin S, PC(u) \leftarrow \infty, HC(u) \leftarrow \infty, PH(u) \leftarrow NULL$; $i \leftarrow i + 1$

3 define link-cost function $C_1(e)$, $e \in E$, with respect to l_w:

$$C_1(e) := \begin{cases} +\infty & \text{if } l_w \text{ traverses link } e\text{, or } \nu_e^* \text{ is equal to } \nu_e^u \text{ for some node } u \\ & \text{along } l_w \text{ and link } e \text{ does not have any free wavelength} \\ \epsilon \cdot C(e) & \text{if for any node } u\ (u \neq s, d) \text{ traversed by } l_w,\ \nu_e^u < \nu_e^* \\ C(e) & \text{otherwise} \end{cases}$$

4 while ($V' \neq \phi$) do {

$u \leftarrow \arg \min_{u \in V'} \{PC(u)\}, V' \leftarrow V' - \{u\}$

if ($u = Head(l_w)$) or (l_w does not traverse u) {

$\forall\, v \in V'$, s.t. $\langle u, v \rangle \in E$

if $PC(v) > PC(u) + C_1(\langle u, v \rangle)$ then {

$PC(v) \leftarrow PC(u) + C_1(\langle u, v \rangle)$

$HC(v) \leftarrow HC(u) + 1$

$PH(v) \leftarrow u$

} // if

} // if

} // while

5 starting from node d and following the reverse direction of l_w, find the first node v which satisfies $HC(v) \leq H_b$

6 retrieve the least-cost path destined to node v by following $PH(v)$ and denote the path as l_b^i

7 allocate backup wavelengths along l_b^i: let l_w^i be the working segment starting from $Head(l_w)$ and ending at v along l_w (inclusively); for any link e that l_b^i traverses and for any node u along l_w^i (excluding the source and the destination nodes of l_w^i), $\nu_e^u \leftarrow \nu_e^u + 1$; if $\nu_e^u > \nu_e^*$, then reserve one more wavelength on link e and let $\nu_e^* \leftarrow \nu_e^u$

8 if v is d, then return $\{l_b^i\}$

9 $S \leftarrow$ all the nodes between $Head(l_w)$ and node v along l_w, excluding $Head(l_w)$ and node v; if S is empty, then undo any changes made to G in Step 7 and return NULL

10 $l_w \leftarrow$ the path starting from node v to node d along l_w

11 go to Step 2

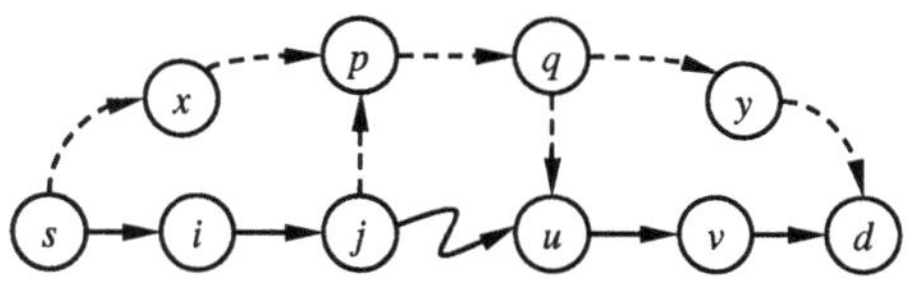

Figure 4.11. Two backup segments, $\langle s, x, p, q, u\rangle$ and $\langle j, p, q, y, d\rangle$, of the same lightpath share the same wavelength link on link $\langle p, q\rangle$, assuming $H_b = 4$.

4) Sometimes, it may be desirable that the hop count of any working segment plus the hop count of its backup segment is no more than some constant H. We can modify Step 5 to cater to this constraint as follows. For any node v along path l_w, denote as $l_b^{i,v}$ the least-cost path destined to node v and denote as h_w^v the number of hops from $Head(l_b^{i,v})$ to node v along l_w. Starting from node d and following the reverse direction of l_w, find the first node v which satisfies the constraint $HC(v) + h_w^v \le H$. Other constraints based on combinations of working and backup segment hop count also can be easily incorporated.

Computational Complexity: The computational complexity of Algorithm 4.2 is $O(|V|^2 + |E|)$. In particular, the computational complexities for Steps 1-11 are $O(1)$, $O(|V|)$, $O(|E|)$, $O(|V|^2 + |E|)$, $O(|V|)$, $O(|V|)$, $O(|V|)$, $O(|E|)$, $O(|V|)$, $O(|V|)$, and $O(1)$, respectively. If we compute K candidate working paths and execute Algorithm 4.2 for each candidate working path, then the computational complexity is $O(K \cdot (|V|^3 + |E|))$.

4.3.3 Illustrative Numerical Results

We compare our GSP_QoP approach to shared-path protection under the same network configuration as described in Section 4.2.3. For the illustrative results shown here, we use $K = 2$ as we found the performance improvement is marginal if we increase K to any larger value.

4.3.3.1 Blocking probability under different values of ϵ

Figure 4.12 plots the blocking performance for $H_b = 6$ under $\epsilon = 0.01, 0.49$, and 0.99. Recall that ϵ is the parameter used by the link-cost function in computing a shared backup path, and smaller values of ϵ lead to better backup sharing. We observe that our GSP_QoP approach has drastically lower blocking probability than shared-path protection under the same ϵ. We further observe that large values of ϵ, e.g., $\epsilon = 0.49$ or $\epsilon = 0.99$, are preferable as both GSP_QoP and shared-path protection have significantly lower blocking when ϵ has a large value. Later, we shall use large values of ϵ.

Figure 4.13 shows the performance gain, defined in Section 4.2.3.2, for $H_b = 6$ under $\epsilon = 0.01, 0.49$, and 0.99. We observe that GSP_QoP has a remarkable performance gain (over 70% across all load regions). The huge

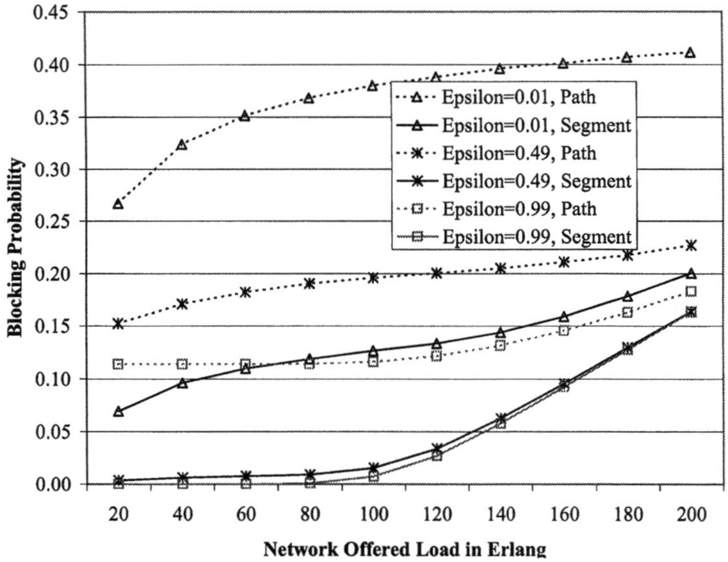

Figure 4.12. Blocking probability for $H_b = 6$.

performance gain results from the fact that GSP_QoP relaxes the path-wise node-/link- disjointness to segment-wise node-/link- disjointness and computes segments with respect to H_b.

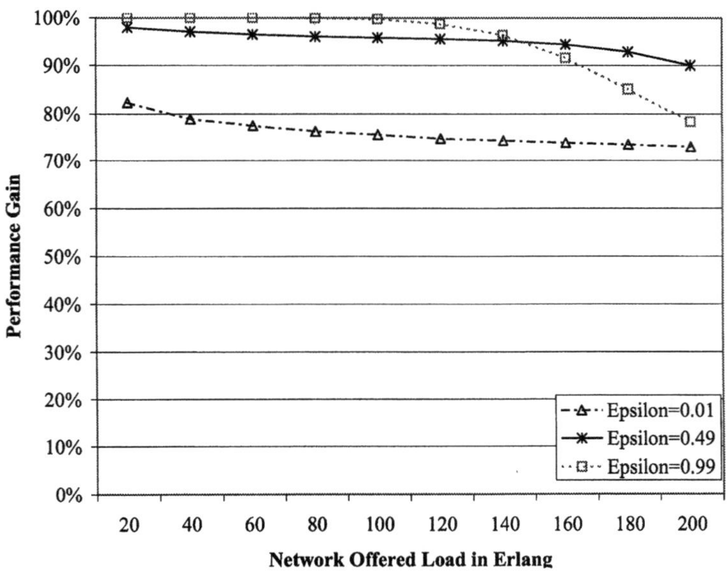

Figure 4.13. Performance gain of GSP_QoP over shared-path protection for $H_b = 6$.

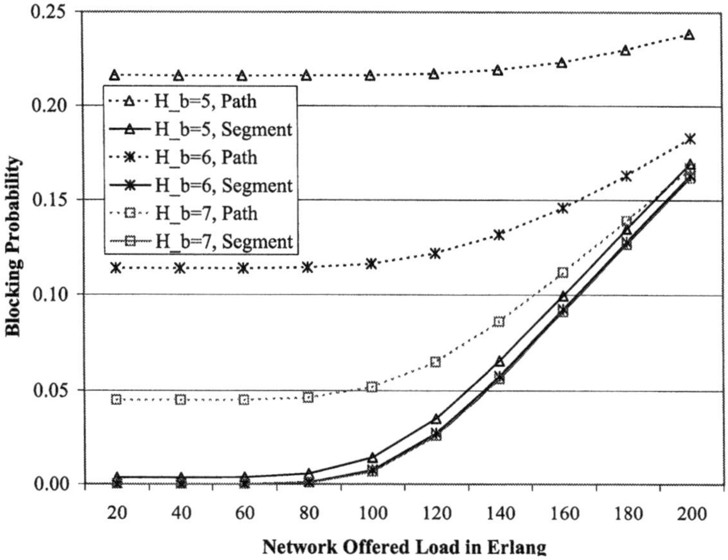

Figure 4.14. Blocking probability for $\epsilon = 0.99$.

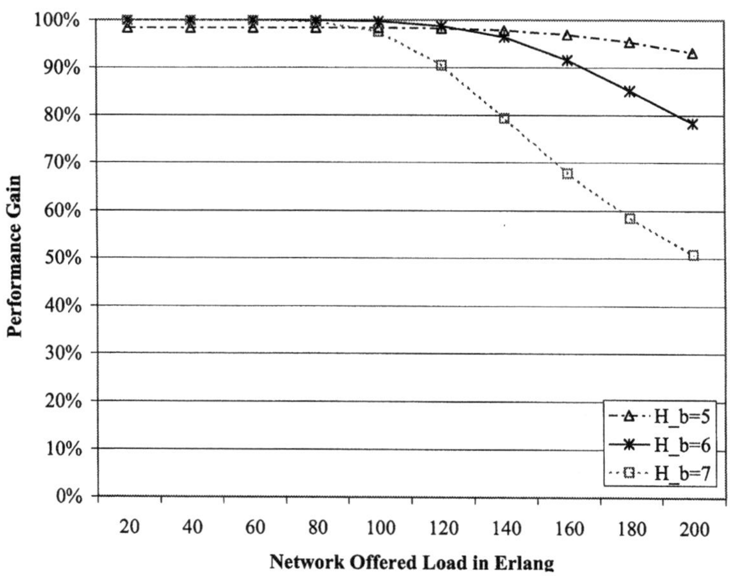

Figure 4.15. Performance gain for $\epsilon = 0.99$.

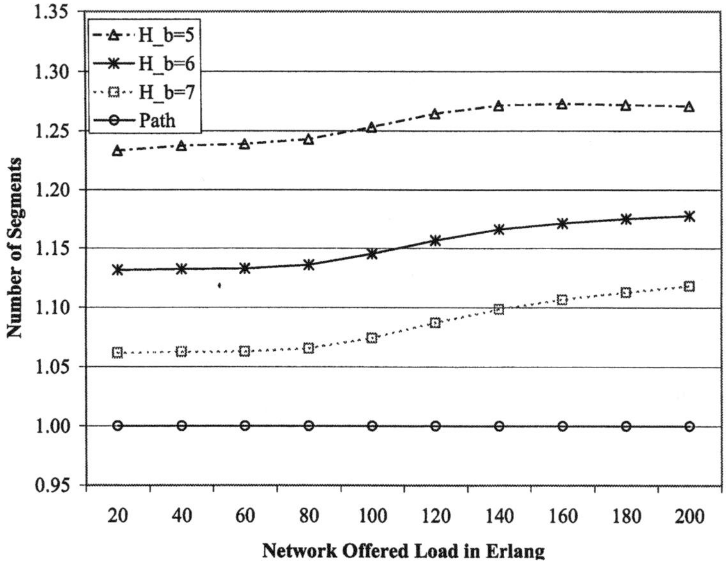

Figure 4.16. Number of segments per lightpath for $\epsilon = 0.99$.

4.3.3.2 Blocking probability under different values of H_b

Figure 4.14 examines the impact of H_b on both GSP_QoP and shared-path protection with $\epsilon = 0.99$. We observe that: (a) GSP_QoP has much lower blocking probability when the load is not very high. (b) When H_b increases from 5 to 6, GSP_QoP has noticeable reduction in blocking probability while the reduction is marginal when H_b further increases to 7. (c) As H_b increases, the blocking probability of shared-path protection drops significantly. However, shared-path protection still has remarkable blocking (above 4%) even when network offered load is low, e.g., 20 Erlangs which translate to about 8.5% average link utilization. This is due to the path-wise end-to-end node-/link-disjoint nature of shared-path protection.

The effectiveness of our GSP_QoP can be further observed in Fig. 4.15 in terms of performance gain. When load is modest or low, GSP_QoP achieves close to 100% performance gain. Even when load is high and wavelengths are heavily used, GSP_QoP still achieves more than 50% performance gain.

Figure 4.16 shows that more segments are needed for smaller value of H_b. However, the average number of segments per lightpath is still quite low, e.g., less than 1.3 even for $H_b = 5$. This implies that the control and management overhead due to segmentation is not very significant.

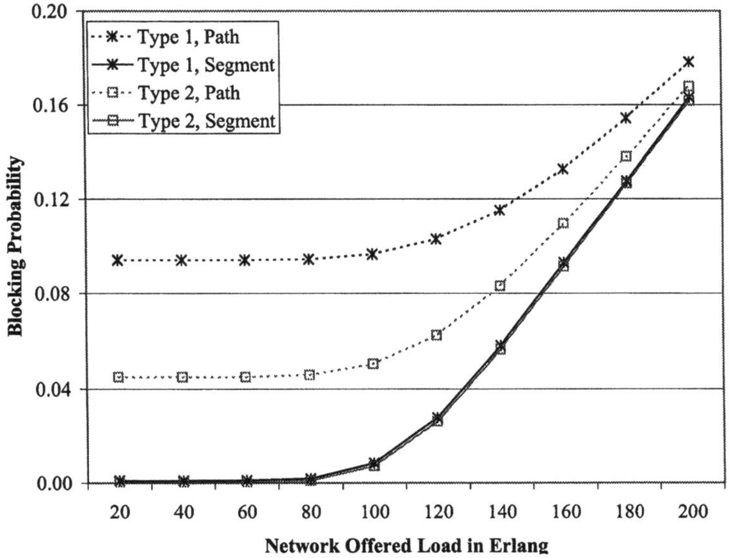

Figure 4.17. Blocking probability ($\epsilon = 0.99$). In Type 1, H_b follows $5 : 6 : 7 : \infty = 30 : 20 : 10 : 40$; In Type 2, H_b follows $5 : 6 : 7 : \infty = 10 : 20 : 20 : 50$.

4.3.3.3 Blocking probability for lightpath requests with differentiated QoP requirements

Different lightpath requests may have different QoP requirements, as discussed earlier in Section 4.3.1. Figures 4.17 and 4.18 compare the performance of GSP_QoP to shared-path protection under two types of traffic. The QoP of the lightpath requests in terms of H_b follows the distribution $5 : 6 : 7 : \infty = 30 : 20 : 10 : 40$ in Type 1 and $5 : 6 : 7 : \infty = 10 : 20 : 20 : 50$ in Type 2.

GSP_QoP has much lower blocking probability than shared-path protection, as shown in Fig. 4.17. For shared-path protection, the large difference between the blocking probability for the two types of traffic implies that shared-path protection cannot effectively provision lightpath requests based on their differentiated QoP requirements. In contrary, the difference between the blocking probability for the two types of traffic in GSP_QoP is very small. This indicates that GSP_QoP can properly provision lightpath requests according to their differentiated QoP requirements.

As shown in Fig. 4.18, when load is modest or low, GSP_QoP achieves close to 100% performance gain; even when load is high, GSP_QoP still achieves more than 35% performance gain.

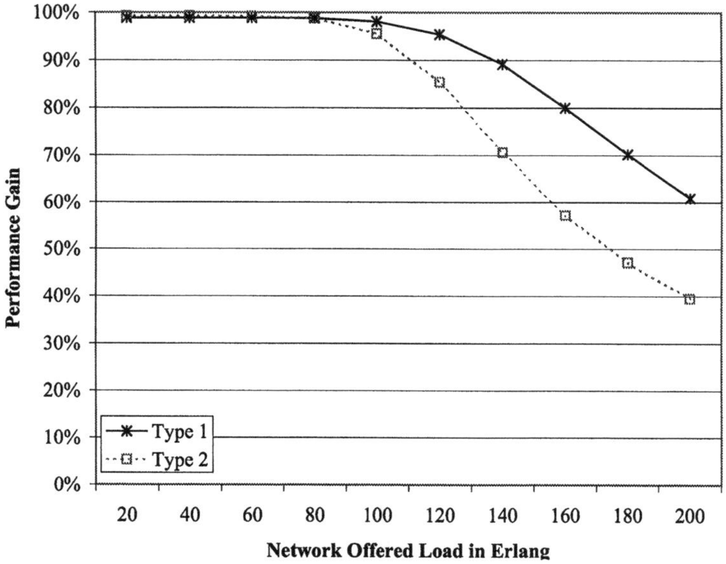

Figure 4.18. Performance gain ($\epsilon = 0.99$). In Type 1, H_b follows $5 : 6 : 7 : \infty = 30 : 20 : 10 : 40$; In Type 2, H_b follows $5 : 6 : 7 : \infty = 10 : 20 : 20 : 50$.

4.3.3.4 Blocking probability for different values of H

Sometimes, it may be desirable that the hop count of any working segment plus the hop count of its backup segment be no more than some constant H. This can be easily incorporated into GSP_QoP as discussed earlier.

Figures 4.19 and 4.20 examines the impact of H on GSP_QoP and shared-path protection. The curves in Figs. 4.19–4.20 have similar trend to the ones in Figs. 4.14–4.15 and can be explained similarly. Meanwhile, since H applies to both working and backup segments, as opposed to H_b which applies only to backup segments, the H constraint is more stringent than the H_b constraint. As a result, the performance gain for different values of H is even higher—namely, above 70% across all load regions—as shown in Fig. 4.20 since segmented protection leads to shorter working and backup segments as shown earlier in Fig. 4.6.

4.4 Conclusion

This chapter considered the problem of dynamic survivable lightpath provisioning against single node/link failures in optical WDM mesh networks. We unified various forms of segment protection into generalized segment protection (GSP). We designed an efficient heuristic which, upon the arrival of a lightpath request, dynamically divides a judiciously-selected working path into multiple overlapped working segments and computes a backup segment for each

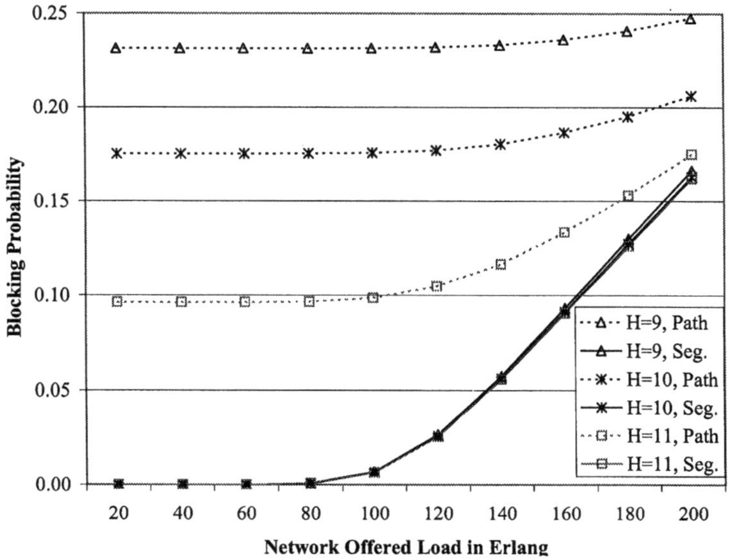

Figure 4.19. Blocking probability ($\epsilon = 0.99$).

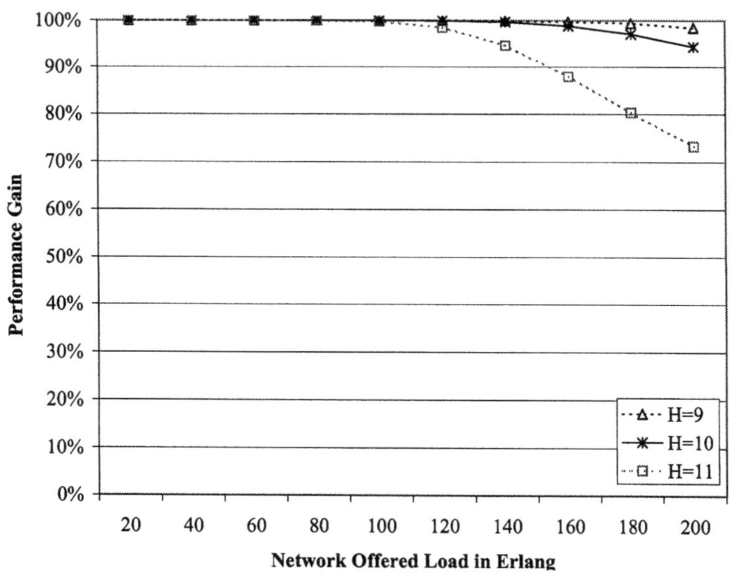

Figure 4.20. Performance gain ($\epsilon = 0.99$).

working segment while accommodating backup sharing. Comparison between GSP and shared-path protection demonstrated that, for a little sacrifice in control and management overhead, GSP achieves much lower blocking probability and shorter protection-switching time.

Based on generalized segment protection, we presented a new approach to provisioning lightpath requests according to their differentiated quality-of-protection (QoP) requirements. We focused on one of the most important QoP parameters—namely, protection-switching time—since lightpath requests may have differentiated protection-switching-time requirements. Numerical results showed that our approach achieves significant performance gain which leads to remarkable reduction in blocking probability.

While our focus is on WDM network, our approaches can also be applied to MPLS networks with appropriate variations, e.g., differentiated bandwidth granularities.

We have considered connections of full wavelength capacity so far. The next chapter considers connections of sub-wavelength granularity.

Chapter 5

SURVIVABLE TRAFFIC GROOMING–DEDICATED PROTECTION

5.1 Introduction

While the transmission rate of a wavelength channel is high (typically STS-192 today and STS-768 expected to be deployed), the bandwidth requirement of a typical connection request can vary from the full wavelength capacity down to STS-1 or lower. To efficiently utilize network resources, sub-wavelength-granularity connections can be groomed onto direct optical transmission channels, or lightpaths[1]. Meanwhile, the failure of a network element can cause the failure of several lightpaths, thereby leading to large data and revenue loss. Fault-management schemes such as protection are essential to survive such failures.

Different low-speed connections may request different bandwidth granularities as well as different protection schemes (dedicated, shared, or no protection). How to efficiently groom such low-speed connections while satisfying their protection requirements is our main focus.

Since some mission-critical services may desire dedicated protection for fast protection switching, we investigate the problem of dynamic sub-wavelength-granularity connection provisioning with dedicated protection against single-

[1] We distinguish the terms "lightpath" and "connection" as follows. The bandwidth requirement of a lightpath is the full wavelength capacity (STS-192 in our present study). The bandwidth requirement of a connection can be any quantized value no more than the full wavelength capacity. Later in our examples and results, we use the quantized values STS-1, STS-3c, STS-12c, STS-48c, and STS-192c for illustration purposes since these values have been widely used in current systems (the "c" after the number implies this is a contiguous block of STS-1s that are part of the same connection). We use the term "STS-n" to refer to the payload carried within an OC-n optical interface ($n = 1, 3, 12$, etc.).

fiber failures under a generic grooming-node architecture in this chapter[2]. We concentrate on single-fiber failures because they are the predominant form of failures in communication networks.

The rest of this chapter is organized as follows. The remainder of this section provides background information. Section 5.2 presents a generic grooming-node architecture. Section 5.3 formally states the problem. Section 5.4 presents two approaches—protection-at-lightpath (PAL) level and protection-at-connection (PAC) level—and provides a qualitative comparison. Sections 5.5 and 5.6 present heuristic algorithms for PAL and PAC. Section 5.7 compares PAL and PAC under different network configurations. Section 5.8 concludes this chapter.

5.1.1 Traffic Grooming

Traffic grooming refers to the problem of efficiently packing low-speed connections onto high-capacity wavelength channels to better utilize network resources [Modiano and Lin, 2001, Zhu and Mukherjee, 2003].

Traffic grooming on SONET/WDM ring networks has been extensively studied; see, for example, [Gerstel et al., 2000, Simmons et al., 1999, Wang et al., 2001, Wang and Mukherjee, 2002, Zhang and Qiao, 2000]. In WDM mesh networks, the traffic-grooming problem has mainly addressed static traffic where a traffic demand matrix is known a priori [Zhu et al., 2003a, Zhu and Mukherjee, 2002b]. On-line approaches for traffic grooming in WDM mesh networks have been recently reported in [Thiagarajan and Somani, 2001a, Zhu et al., 2002, Zhu and Mukherjee, 2002a, Zhu et al., 2003b]. The work in [Thiagarajan and Somani, 2001a] proposes a call-admission-control algorithm to address the capacity-fairness issue, i.e., a connection request with higher bandwidth requirement is more likely to be blocked than a connection request with lower bandwidth requirement. The work in [Zhu and Mukherjee, 2002a] proposes different grooming policies and route-computation algorithms for different network states. The work in [Zhu et al., 2002, Zhu et al., 2003b] develops an algorithm for dynamically grooming low-speed connections to meet different traffic-engineering objectives based on the generic graph model proposed in [Zhu et al., 2003a]. Please see [Zhu and Mukherjee, 2003] for an extensive review on traffic grooming.

5.1.2 Lightpath Protection

A significant amount of research has been done on dynamic, survivable lightpath provisioning. We briefly review a portion of the related work [Ho

[2] In the next chapter, we explore the problem of dynamic low-speed connection provisioning with shared protection against single-fiber failures. We remark that these two problems are significantly different in terms of treatment. Thus, each problem and its findings are documented separately.

and Mouftah, 2002, Ramamurthy et al., 2001, Zhang et al., 2003], focusing on dedicated protection. (Please see [Ellinas et al., 2003] for an extensive overview.) The work in [Ramamurthy et al., 2001] describes an architecture for dynamic lightpath provisioning and analyzes the performance for dynamically-provisioned unprotected, 1+1 protected, and shared-mesh-protected lightpaths. The work in [Ho and Mouftah, 2002] presents short leap shared protection (SLSP), the basic idea of which is to divide a working path into overlapped segments and protect each segment individually. The work in [Zhang et al., 2003] proves the problem of jointly computing two link-disjoint paths in a wavelength-continuous network to be $\mathcal{NP}$-complete and presents different heuristic algorithms. The work in [Kar et al., 2002] presents two algorithms for dynamically routing dedicated-protected connections in a load-balanced manner.

5.1.3 Survivable Traffic Grooming

The survivable traffic-grooming problem, in which sub-wavelength-granularity connections need to be protected, is a relatively unexplored territory.

Given a static traffic matrix and the protection requirement of each connection request, the work in [Lardies et al., 2001] presents an integer linear program and a heuristic for satisfying the bandwidth and protection requirements of all the connection requests while minimizing the network cost in terms of transmission cost and switching cost.

For dynamically establishing low-speed connection requests with shared protection, the work in [Thiagarajan and Somani, 2001b] presents mixed working-backup grooming policy (MGP) and segregated working-backup grooming policy (SGP). With both schemes employing fixed-alternate routing [Ramamurthy and Mukherjee, 2002], the work focuses on the effect of different wavelength-assignment algorithms and different topologies.

5.1.4 Our Proposal

In wavelength-convertible networks, while it may be straightforward to dynamically provision a lightpath request with dedicated protection by applying Suurballe's algorithm [Suurballe and Tarjan, 1984] or its variant, the Bhandari's algorithm [Bhandari, 1999], the introduction of grooming constraints (shown in Section 5.2) increases the problem complexity significantly. In fact, we shall show in Section 5.6 that the problem complexity increases from $\mathcal{P}$ to $\mathcal{NP}$-complete (from computational-complexity point of view). Thus, fast and practical heuristics are needed.

We propose two approaches—protection-at-lightpath (PAL) level and protection-at-connection (PAC) level—for dynamically provisioning dedicated-protected low-speed connection requests against single-fiber failures. We investigate their characteristics under a generic grooming-node architecture, and we design efficient heuristics. Our work differs from previous work in that we focus on route computation and the impact of different resource constraints such as wavelength and grooming capacity.

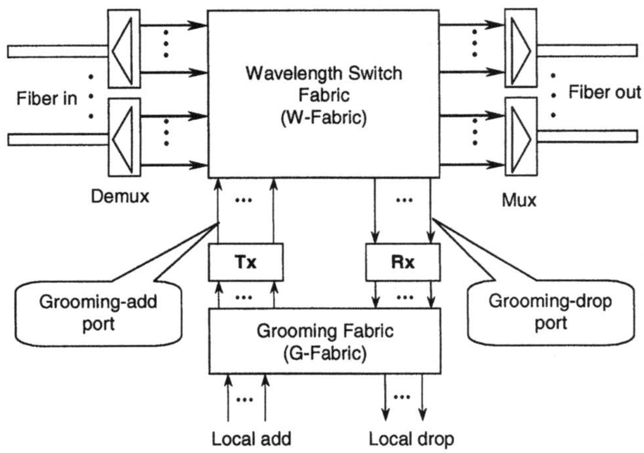

Figure 5.1. A simplified grooming-node architecture.

5.2 Grooming-Node Architecture

In order to support traffic grooming, a network node should be able to switch traffic at wavelength granularity and finer granularity. Figure 5.1 shows the logical view of a simplified grooming-node architecture.

This hierarchical grooming node consists of a wavelength-switch fabric (W-Fabric) and a grooming fabric (G-Fabric). The W-Fabric performs wavelength routing; the G-Fabric performs multiplexing, demultiplexing, and switching of low-speed connections. A portion of the incoming wavelengths to the W-Fabric can be dropped to the G-Fabric through the grooming-drop ports for sub-wavelength-granularity switching. The groomed traffic can then be added to the W-Fabric through the grooming-add ports. The number of grooming ports determines the grooming capacity of a node (we assume that there are equal number of grooming-add and grooming-drop ports). Later, we shall investigate the impact of grooming capacity (number of grooming ports) on the network performance.

Even though crossconnects capable of full grooming–i.e., G-Fabrics–are preferable to network operators today, crossconnects capable of wavelength switching–i.e., W-Fabrics–are expected to be desirable as traffic continues to grow in the future. The G-Fabrics deployed today are unlikely to go away when W-Fabrics are deployed due to economic reasons. One way of effectively utilizing both G-Fabrics and W-Fabrics could be to interconnect a W-Fabric and a G-Fabric through transponders, as shown in Fig. 5.1.

As a special case, if the number of grooming ports at a node is equal to the number of incoming wavelengths to its W-Fabric, then this grooming node can switch the entire incoming traffic at STS-1 level, as is the case in today's state-of-the-art opaque (i.e., switching with optical-to-electronic-to-optical (OEO) conversion) intelligent optical switches from many vendors.

While our approaches apply to both wavelength-continuous and wavelength-convertible networks, we hereafter assume without loss of generality that the network has full wavelength-conversion capability.

5.3 Problem Statement

We first define the notations and then formally state the dynamic connection-provisioning problem. A network is represented as a weighted, directed graph $G = (V, E, C, \lambda, P)$, where V is the set of nodes, E is the set of unidirectional fibers (referred to as links), $C : E \rightarrow R^+$ is the cost function for each link (where R^+ denotes the set of positive real numbers), $\lambda : E \rightarrow Z^+$ specifies the number of wavelengths on each link (where Z^+ denotes the set of positive integers), and $P : V \rightarrow Z^+$ specifies the number of grooming ports at each node.

A connection request is represented as a quadruple $\langle s, d, B, t_h \rangle$, which specifies the source node, the destination node, the bandwidth requirement, and the holding time in this order. In this chapter, a connection needs dedicated protection (1+1 or 1:1). Thus, the cost of a working (or backup) path is the sum of the cost of the links that path traverses.

We now formally state the dynamic connection-provisioning problem as follows: Given the current network state (which includes the network topology as a weighted digraph G, existing lightpath/connection information [e.g., routes and wavelengths, etc.], wavelength usage, and grooming-port usage), route each connection request with respect to its bandwidth and protection requirement (dedicated protection) while minimizing the incremental cost in terms of the total cost of the working and backup paths under the assumptions that existing connections cannot be disturbed and no information about future arrivals is available at the time of provisioning the current connection.

5.4 Proposed Approaches

To provision a connection request with dedicated protection, there are two types of resource constraints—wavelengths and grooming ports. Typically, the more the number of wavelengths the network has, the less the number of grooming ports a node needs, and vice versa. We propose two schemes—protection-at-lightpath (PAL) level and protection-at-connection (PAC) level—for dedicated protection to explore the tradeoff between wavelengths and grooming ports.

We illustrate PAL and PAC via examples. For the initial network configuration shown in Fig. 5.2, each fiber has one wavelength of capacity STS-192; every node has three grooming ports; and T and R represent the number of available grooming-add and grooming-drop ports, respectively.

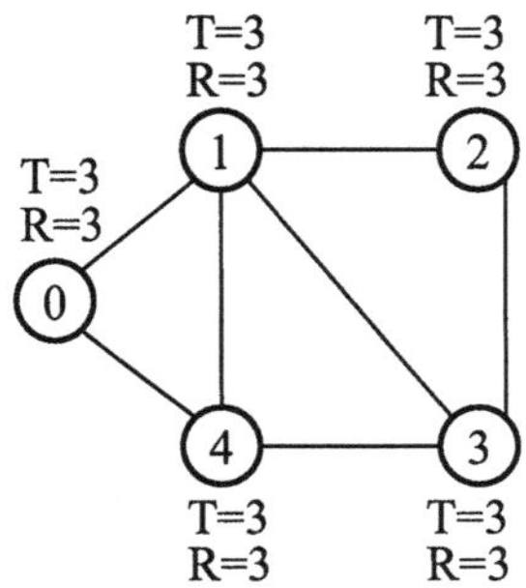

Figure 5.2. Example: initial network configuration (T and R represent the number of free grooming-add and grooming-drop ports at a node, respectively, each with initial value of 3).

5.4.1 Protection-at-Lightpath (PAL) Level

5.4.1.1 Basic idea

PAL provides end-to-end protection with respect to a lightpath. Under PAL, a connection is routed through a sequence of p-lightpaths. A *p-lightpath* is defined as a pair of link-disjoint lightpaths between two nodes. For example, in Fig. 5.3(a), the two link-disjoint lightpaths l_1^w and l_1^b compose p-lightpath l_1. Conceptually, we can view a p-lightpath as a protected lightpath. A connection traversing a sequence of p-lightpaths automatically survives from single-link failures since each p-lightpath survives from single-link failures by definition. Since protection occurs at lightpath level, PAL has the advantages of low implementation complexity and low signaling overhead when a failure occurs.

5.4.1.2 Example

Upon the arrival of the first connection request c_1 ($\langle 4, 1, \text{STS-48c}, t_h \rangle$), one way of provisioning c_1 under PAL is shown in Fig. 5.3(a). Connection c_1 is routed via the p-lightpath l_1, which consists of two link-disjoint lightpaths l_1^w (working) and l_1^b (backup). Both lightpaths l_1^w and l_1^b consume a grooming-add port at node 4 and a grooming-drop port at node 1. The remaining capacity on p-lightpath l_1 is STS-144.

Suppose that c_1 remains in the network when the second connection request c_2 ($\langle 4,\ 2,\ \text{STS-12c},\ t_h' \rangle$) arrives. Based on the current network state, one possible solution is that PAL grooms c_2 to the existing p-lightpath l_1, and it also sets up a new p-lightpath l_2 (with l_2^w as working lightpath and l_2^b as

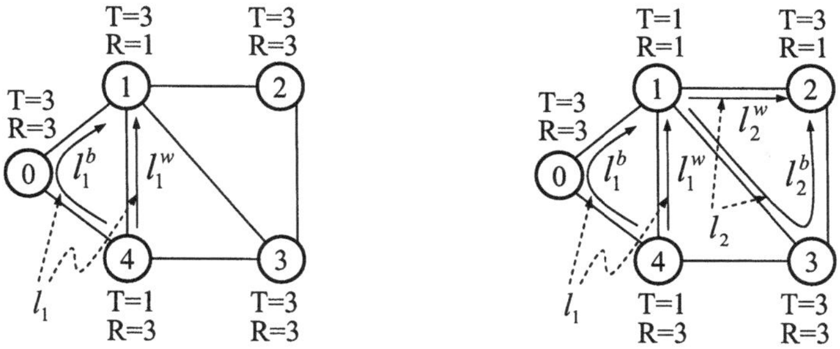

(a) After provisioning c_1 (b) After provisioning c_2

Figure 5.3. PAL: provisioning connections c_1 $(\langle 4, 1, \text{STS-48c}, t_h\rangle)$ and c_2 $(\langle 4, 2, \text{STS-12c}, t_h^{'}\rangle)$.

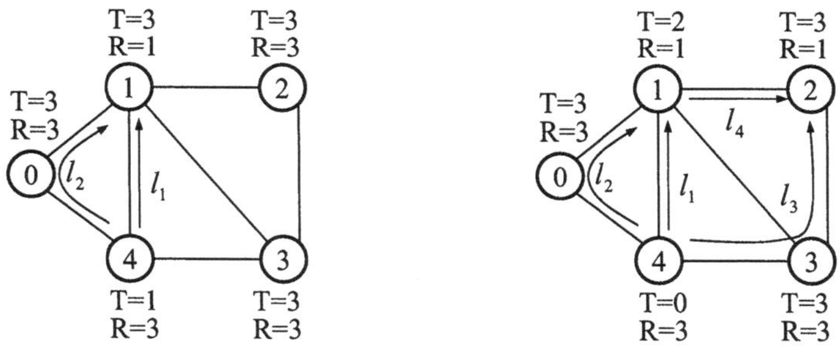

(a) After provisioning c_1 (b) After provisioning c_2

Figure 5.4. PAC: provisioning connections c_1 $(\langle 4, 1, \text{STS-48c}, t_h\rangle)$ and c_2 $(\langle 4, 2, \text{STS-12c}, t_h^{'}\rangle)$.

backup lightpath) as shown in Fig. 5.3(b). Both lightpaths l_2^w and l_2^b consume a grooming-add port at node 1 and a grooming-drop port at node 2. Now, the remaining capacity of l_1 is STS-132 and the remaining capacity of l_2 is STS-180.

An immediate observation is that, in PAL, working (backup) traffic is groomed onto the working (backup) lightpaths of p-lightpaths.

5.4.2 Protection-at-Connection (PAC) Level

5.4.2.1 Basic idea

PAC provides end-to-end protection with respect to a connection. Under PAC, a connection is routed via link-disjoint working and backup paths, each of which traverses a sequence of lightpaths. PAC deals with individual connections and is expected to pack connections efficiently. To illustrate PAC, let us consider the same network topology and connection requests as earlier.

5.4.2.2 Example

When the first connection request c_1 arrives, one way of provisioning c_1 under PAC is shown in Fig. 5.4(a). Two lightpaths l_1 and l_2 have been set up. Connection c_1's primary path traverses lightpath l_1, and its backup path traverses lightpath l_2. Both lightpaths l_1 and l_2 consume a grooming-add port at node 4 and a grooming-drop port at node 1. The remaining capacities on lightpaths l_1 and l_2 are both STS-144. The difference between PAC and PAL after provisioning c_1 is that the full lightpath capacity of l_1^b is reserved as protection resource for lightpath l_1^w in PAL while, in PAC, only a fraction (STS-48) of lightpath l_2's entire bandwidth (STS-192) is reserved to protect a fraction (STS-48) of lightpath l_1's entire bandwidth.

Suppose that connection c_1 remains in the network when the second connection request c_2 arrives. One way of provisioning c_2 under PAC is shown in Fig. 5.4(b). Two more lightpaths l_3 and l_4 have been set up. Lightpath l_3 consumes a grooming-add port at node 4 and a grooming-drop port at node 2. Lightpath l_4 consumes a grooming-add port at node 1 and a grooming-drop port at node 2. The working path for c_2 is lightpath l_3 and the backup path is the two-lightpath sequence $\langle l_1, l_4 \rangle$. Now, the remaining capacity on lightpath l_1 is STS-132; the remaining capacity on lightpath l_2 is STS-144; and the remaining capacity on lightpaths l_3 and l_4 is STS-180.

Clearly, in PAC, working traffic and backup traffic can be groomed onto the same lightpath.

5.4.3 PAL vs. PAC: A Qualitative Comparison

The above illustrative examples indicate that PAL and PAC perform differently in terms of routing and the amount of resources required. Below, we qualitatively compare their characteristics with respect to routing, solution space, and operational complexity.

5.4.3.1 Routing

The fundamental routing difference between PAL and PAC is that PAL provides end-to-end protection with respect to lightpath while PAC provides end-to-end protection with respect to connection. Under PAL, when a failure occurs, the end nodes of the affected p-lightpaths switch to their backup lightpaths; and the affected connections are oblivious to the protection-switching process. Under PAC, when a failure occurs, the end-nodes of the affected connections (which could be significantly more than the number of affected lightpaths) switch to their backup paths.

In this sense, the difference between PAL and PAC is similar to the difference between sub-path protection [Ho and Mouftah, 2002, Ou et al., 2002] (or link protection as one extreme) and path protection with respect to the restoration process after a fault occurs. However, the introduction of grooming constraints and the fact that connections have differentiated bandwidth requirements increase the route-computation complexity significantly.

PAC deals with connections and therefore can pack connections more efficiently than PAL. Furthermore, the two lightpaths of a p-lightpath are an integrated unit and they cannot be utilized individually. As a result, under PAC, low-speed connections are more likely to be groomed onto lightpaths, and more grooming ports are more likely to be consumed. Basically, PAC trades grooming ports for bandwidth efficiency in routing, while PAL trades the bandwidth efficiency in routing each connection for the savings in grooming ports.

5.4.3.2 Solution space

One may think that PAL is a special case[3] of PAC since, for a connection, the concatenation of the working (backup) lightpaths of the p-lightpaths which this connection traverses under PAL may consist of a valid working (backup) path under PAC. The two paths so formed, however, may not be valid for PAC, in general.

Consider the following example. A connection from node s to node d traverses p-lightpaths l_1, l_2, and l_3 (their corresponding working and backup lightpaths are l_i^w and l_i^b, $i = 1, 2, 3$). Suppose that lightpath l_1^w traverses link e_1, $\langle u_1, v_1 \rangle$; lightpath l_1^b traverses link e_2, $\langle u_2, v_2 \rangle$; lightpath l_2^w traverses links e_1 and e_2; lightpath l_2^b traverses links e_3, $\langle u_3, v_3 \rangle$, and e_4, $\langle u_4, v_4 \rangle$; lightpath l_3^w traverses link e_3; and lightpath l_3^b traverses link e_4. Figures 5.5(a) and (b) show the logical and physical routing of the p-lightpaths l_1, l_2, and l_3.

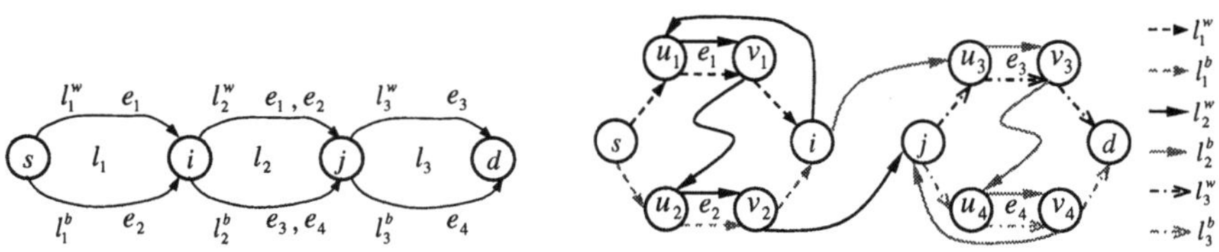

(a) A logical view of p-lightpaths l_1, l_2, and l_3.(b) Physical routing of p-lightpaths l_1, l_2, and l_3.

Figure 5.5. A solution of PAL does not form a solution of PAC.

Then, the path $\langle l_1, l_2, l_3 \rangle$ is a valid solution for this connection under PAL. However, there is no end-to-end (from node s to node d) link-disjoint path pair between node s and node d based on the set of lightpaths $\{l_1^w, l_1^b, l_2^w, l_2^b, l_3^w, l_3^b\}$. This is because lightpaths l_2^w and l_1^w (lightpaths l_2^w and l_1^b, lightpaths l_2^b and l_3^w, as well as lightpaths l_2^b and l_3^b) are in the same shared-risk link group (SRLG) [Strand et al., 2001, Zang et al., 2003]. While paths $\langle s, u_1, v_1, i, u_3, v_3, d \rangle$

[3]Special case in the sense that any valid solution under PAL is a valid solution under PAC.

and $\langle s, u_2, v_2, j, u_4, v_4, d\rangle$ are link-disjoint from node s to node d, we may not always have the freedom to construct the two paths for the following reasons. First, reshuffling the wavelength links utilized by the three p-lightpaths (l_1, l_2, and l_3) may disturb existing traffic since the three p-lightpaths can be set up during different time period and some of them may be carrying live traffic. Second, in a wavelength-continuous network, the two paths so formed may not be wavelength continuous because path segments $\langle i, u_3, v_3\rangle$ and $\langle v_3, d\rangle$ (as well as path segments $\langle s, u_2, v_2\rangle$ and $\langle v_2, j\rangle$) can be on different wavelengths under PAL. Thus, the set of lightpaths $\{l_1^w, l_1^b, l_2^w, l_2^b, l_3^w, l_3^b\}$ cannot form a valid solution under PAC, in general.

5.4.3.3 Operational complexity

From implementation point of view, PAL is simpler to implement than PAC. To provision a dedicated-protected connection request, PAL does not need the routing information of any existing lightpaths. PAC, however, does need the routing information of all the existing lightpaths. Essentially, PAL performs at an aggregate level (lightpath) and PAC works on a per-flow basis (connection).

In the case of 1:1 dedicated protection, PAL has lower signaling overhead from control point of view. Assume that a lightpath can carry up to g connections. (In today's networks, g is typically 192 since wavelength capacity is STS-192 and the lowest bandwidth granularity is typically STS-1.) When a link fails, W lightpaths can be disrupted in the worst case. In PAL, at most W protection-switching processes are needed. However, in PAC, up to $W \times g$ protection-switching processes are required in the worst case. As protection-switching processes for 1:1 dedicated protection typically require signaling, PAL will consume lower control bandwidth and involves lower signaling complexity compared to PAC.

5.5 PAL Heuristic

5.5.1 Problem Complexity

Let us consider the problem of deciding whether there exists a valid route under PAL for a connection request. This problem can be formally stated as follows: Given the topology graph $G = (V, E, C, \lambda, T)$, existing lightpath routing and wavelength-assignment information, wavelength usage, grooming-port usage, and a connection request $\langle s, d, B, t_h\rangle$, does there exist from node s to node d a path consisting of p-lightpaths (existing or new) having sufficient free capacity? We conjecture that this problem is NP-complete. The intuitive explanation follows. If there does not exist such a path utilizing only existing p-lightpaths or only free wavelength links, then such a path (if it exists) needs to utilize some existing p-lightpaths and some free wavelength links (which should be able to form some p-lightpaths). To decide which p-lightpath and which free

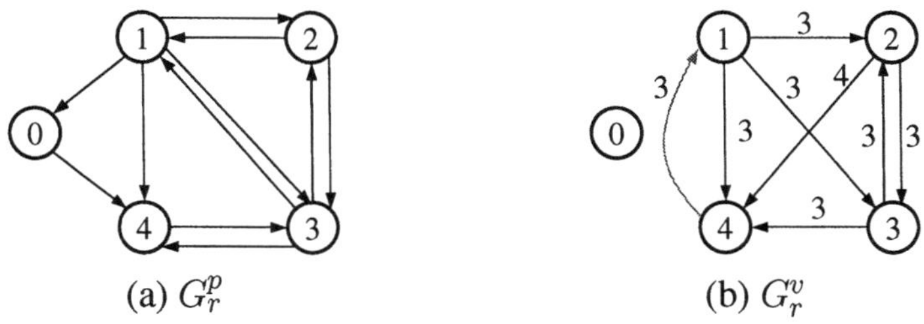

Figure 5.6. PAL: (a) physical reachability graph G_r^p, (b) virtual reachability graph G_r^v for the network state in Fig. 5.3(a).

wavelength link to use such that the newly employed wavelength links form some p-lightpaths and these p-lightpaths combined with the previously decided p-lightpaths should form a path, an algorithm needs to enumerate all the possible combinations. The complexity of enumerating all possible link combinations is $O(2^{|E|})$. Thus, we resort to a heuristic in the following subsection.

5.5.2 PAL Heuristic

The basic idea of our PAL heuristic is to construct a virtual reachability graph, in which the vertex set is the same as the original graph and there exists a link from node u to node v if node v is reachable from node u through either an existing p-lightpath of sufficient free capacity or a new p-lightpath subject to grooming-port and wavelength constraints. We can then apply a shortest-path algorithm on the virtual reachability graph to compute a sequence of p-lightpaths. The PAL heuristic is shown in Algorithm 5.1.

5.5.3 Explanation

Using the same example as in Section 5.4.1, we illustrate how connection c_2 is provisioned by the PAL heuristic. Suppose the network state after provisioning connection c_1 is the same as in Fig. 5.3(a) and the cost of each link is unity. When connection c_2 ($\langle 4, 2, \text{STS-12c}, t_h' \rangle$) arrives, PAL constructs the physical reachability graph G_r^p shown in Fig. 5.6(a), according to the current wavelength availability. Please note that links $\langle 0, 1 \rangle$, $\langle 4, 0 \rangle$, and $\langle 4, 1 \rangle$ are gone as they have been utilized by p-lightpath l_1.

In the next step, PAL constructs the virtual reachability graph G_r^v shown in Fig. 5.6(b), according to the available grooming ports, wavelengths, and free capacity of existing p-lightpath. In Fig. 5.6(b), the number on a link is its cost, which is computed in Step 2 of Algorithm 5.1. Among the eight links in this figure, link $\langle 4, 1 \rangle$ is the only one corresponding to an existing p-lightpath (l_1 in Fig. 5.3(b)), while the other links corresponding to two link-disjoint paths in the physical reachability graph G_r^p (e.g., link $\langle 1, 2 \rangle$ in G_r^v corresponds to two link-disjoint paths $\langle 1, 2 \rangle$ and $\langle 1, 3, 2 \rangle$ in G_r^p).

Algorithm 5.1 PAL

Input: $G = (V, E, C, \lambda, P)$, $c = \langle s, d, B, t_h \rangle$, existing p-lightpath information, wavelength usage, and grooming-port usage.
Output: A survivable route (a sequence of p-lightpaths), or NULL if no such route is found.

1 Construct a *physical reachability graph* $G_r^p = (V, E_r^p, C)$, where, for any $u, v \in V$, $\langle u, v \rangle \in E_r^p$ if there exists a free wavelength from node u to node v in G. The cost of $\langle u, v \rangle$ in G_r^p is the same as its cost in G.

2 Construct a *virtual reachability graph* $G_r^v = (V, E_r^v, C_r^v)$, where, for any $u, v \in V$, $\langle u, v \rangle \in E_r^v$ if there exists an existing p-lightpath (from u to v) of free capacity no less than B, or if a new p-lightpath from u to v can be set up subject to grooming-port and wavelength constraints. Specifically, let L_{uv} be the set of existing p-lightpaths from node u to node v of free capacity no less than B. Let p-lightpath l_{uv}^e be NULL if L_{uv} is empty; otherwise, let l_{uv}^e be the p-lightpath of minimal cost among all such p-lightpaths in L_{uv}. Let p-lightpath l_{uv}^n be NULL if node u has less than two free grooming-add ports, or node v has less than two free grooming-drop ports, or no link-disjoint paths from u to v exist in G_r^p; otherwise, let l_{uv}^n be the link-disjoint path pair of minimal cost from u to v. Link $\langle u, v \rangle \in E_r^v$ if either l_{uv}^e or l_{uv}^n is not NULL, and $\langle u, v \rangle$ is associated to the p-lightpath l_{uv}^e (l_{uv}^n) if l_{uv}^e (l_{uv}^n) has less cost. The cost of $\langle u, v \rangle$ in G_r^v is the cost of the p-lightpath it is associated to. The cost of a p-lightpath is the total cost of its working and backup paths.

3 Apply a shortest-path algorithm on G_r^v from node s to node d. Return NULL if no such path exists; otherwise, let l_s be a shortest path.

4 Compute the number of free wavelengths needed on each physical link based on the set of new p-lightpaths l_s traverses. (Please note that grooming constraints—grooming-port and capacity—have been accommodated in the construction of G_r^v.) Return l_s (and update resource-usage information) if every physical link has sufficient number of free wavelengths; otherwise, return NULL. (There might be some rare situation in which the heuristic cannot find a route, while a valid route exists. Such a situation may occur only in this step. The scenario is that PAL finds a shortest path in G_r^v, but multiple p-lightpaths along the shortest path need to be set up and they contend for a wavelength on some link. The main reason why such a situation is rare is that the lightpath-hop distance of a route is small, and every lightpath hop proceeds towards the destination node. In fact, in all our simulation experiments, such a situation never occurred.)

Afterwards, PAL computes from node 4 to node 2 a shortest path, which turns out to be $\langle 4, 1, 2 \rangle$, in the virtual reachability graph G_r^v. Then, PAL checks if there are sufficient number of free wavelengths on each physical link. For this shortest path $\langle 4, 1, 2 \rangle$, only the second hop $\langle 1, 2 \rangle$ (which corresponds to paths $\langle 1, 2 \rangle$ and $\langle 1, 3, 2 \rangle$ in G_r^p) needs one wavelength along links $\langle 1, 2 \rangle$, $\langle 1, 3 \rangle$, and $\langle 3, 2 \rangle$. As there are enough number of free wavelengths on these links, the connection is accepted.

5.5.4 Optimality

How optimal is the route computed by PAL? We make the following claim to answer this question.

CLAIM 2 *If PAL returns a non-NULL route, then the route is optimal (in the sense that the total cost of the links the route traverses is minimal among all possible routes).*

Proof: Suppose that PAL returns a non-NULL route $\mathcal{R}$. Assume there is another valid route $\mathcal{R}'$ of cost less than that of $\mathcal{R}$. By the construction of the virtual reachability graph $G_r^v = (V, E_r^v, C_r^v)$, for any two nodes $u, v \in V$, if $\langle u, v\rangle \in E_r^v$, then $\langle u, v\rangle$ corresponds to the p-lightpath of minimal cost among all the possible p-lightpaths from node u to node v of sufficient capacity; if $\langle u, v\rangle \notin E_r^v$, then no p-lightpath of sufficient capacity can be found or set up from node u to node v. If we map each p-lightpath which $\mathcal{R}'$ traverses to a link in G_r^v subject to the constraint that the p-lightpath and the link have the same source-destination pair (the cost of the link is no more than the cost of the p-lightpath by the definition of G_r^v), then route $\mathcal{R}'$ corresponds to a path in G_r^v. This contradicts that $\mathcal{R}$ is a minimal-cost path in G_r^v. ∎

5.5.5 Variations

Some variations of Algorithm 5.1 are possible and may be desirable for certain connection-bandwidth distributions. For example, if lower-bandwidth connections significantly outnumber higher-bandwidth connections, then STS-192 connections are more likely to be blocked if Algorithm 5.1 is directly applied. This is because the lower-bandwidth connections may set up many lightly-loaded lightpaths and quickly consume the available wavelengths. To address this capacity-fairness issue, we redefine the cost of an existing p-lightpath as the sum of the cost of the underlying links which the p-lightpath traverses minus a cost-slack parameter δ (if the resultant cost is negative, redefine it as an infinitesimal constant such as 10^{-6}). This will encourage the lower-bandwidth connections to use existing p-lightpaths.

5.5.6 Computational Complexity

The complexity of Algorithm 5.1 is $O(|V|^4)$. In particular, the complexity of Step 1 is $O(|V| + |E|)$; the complexity of Step 2 is $O(|V|^4)$ as PAL applies Suurballe's algorithm for each node pair; the complexity of Step 3 is $O(|V|^2)$; and the complexity of Step 4 is $O(|E|)$.

5.6 PAC Heuristic

We first prove that the existence version of routing a connection request according to PAC under resource constraints is $\mathcal{NP}$-complete in Theorem 5.1.

This problem, referred to as WDM-PAC, can be formally stated as follows: Given the topology graph $G = (V, E, C, \lambda, P)$, the lightpath database, wavelength usage, grooming-port usage, and a connection request $\langle s, d, B, t_h \rangle$, do there exist from node s to node d at least two end-to-end physically link-disjoint paths of capacity no less than B (subject to the resource constraints)?

THEOREM 5.1 *WDM-PAC is $\mathcal{NP}$-complete.*

Proof: Please see Appendix 5.A. ■

Since the problem is $\mathcal{NP}$-complete, we resort to a heuristic. Our PAC heuristic computes two link-disjoint paths based on the current network state.

5.6.1 Grooming-Node Modeling and Network-State Representation

Under the current network state, a connection request may be carried by existing lightpaths, by newly-established lightpaths (based on available wavelengths and free grooming ports), or by both existing lightpaths and newly-setup lightpaths. While the graph defined in Section 5.3 takes into account wavelength constraints, the graph does not accommodate existing lightpath information. Moreover, grooming-port constraints apply if a connection is to be carried by both existing lightpaths and newly-established lightpaths. Therefore, a more powerful mechanism—which can accommodate wavelength constraints, grooming-port constraints, and existing lightpath information—is needed to represent the network state and to facilitate route computation.

We adopt the generic graph model in [Zhu et al., 2003a] to represent the network state as an auxiliary graph. For our grooming-node architecture in Fig. 5.1, W-Fabric is modeled as the λ layer consisting of input vertex[4] λ_I and output vertex λ_O; G-Fabric is modeled as the access layer consisting of input vertex A_I and output vertex A_O; grooming-add port is modeled by an edge from vertex A_O to vertex λ_O; and grooming-drop port is modeled by an edge from vertex λ_I to vertex A_I. A unidirectional fiber is represented as an edge from vertex λ_O at the source node to vertex λ_I at the destination node of the lightpath. A lightpath layer, consisting of input vertex L_I and output vertex L_O, is added to model existing lightpaths sourced/sunk at a node. A lightpath is represented as an edge from vertex L_O at the source node to vertex L_I at the destination node. Every edge is associated with two attributes: one indicating the available capacity and the other indicating the cost of the resource which the edge represents.

As an example, the state of node 4 in Fig. 5.4(b) is modeled in Fig. 5.7. For the four auxiliary edges—$\langle \lambda_I, \lambda_O \rangle$, $\langle L_I, A_I \rangle$, $\langle A_I, A_O \rangle$, and $\langle A_O, L_O \rangle$—the

[4]For clarity, we refer to node and link in the auxiliary graph as vertex and edge.

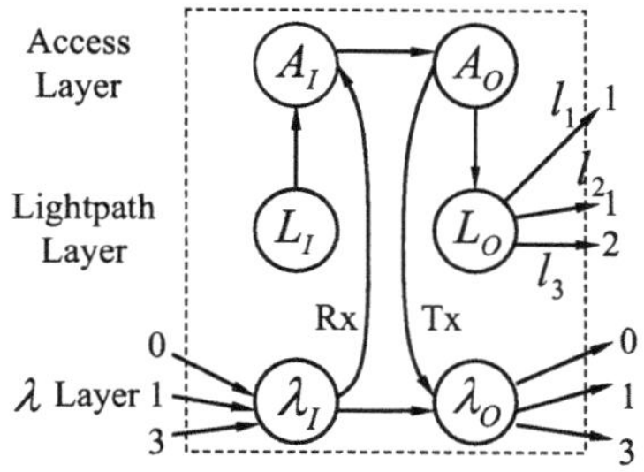

Figure 5.7. Modelling of node 4 in Fig. 5.4(a).

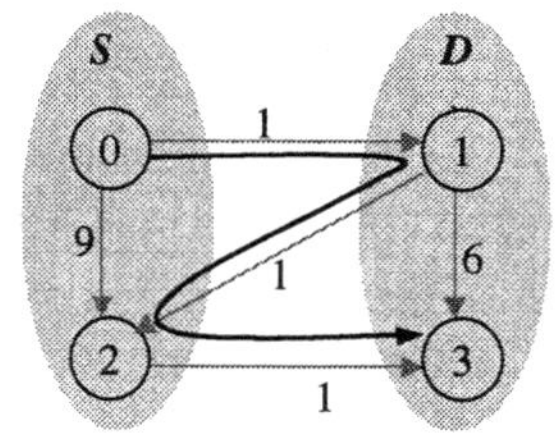

Figure 5.8. Overcoming a "trap" topology.

capacity is infinity and the cost is zero. The available capacity of any other edge e is the available capacity of the resource which edge e represents, e.g., the free capacity of edge $\langle A_O, \lambda_O \rangle$ is zero since $T = 0$ for node 4. The cost of any other edge e is the cost of the resource which edge e represents, e.g., the cost of a lightpath edge is the sum of the cost of the links which the lightpath traverses. The cost of lightpath edge l_2 is two if we assume unity link cost (multiple edges between the same vertex pair are distinguished by unique sequence numbers).

By modeling every grooming node as above, the current network state—which includes wavelength usage, grooming-port usage, and available lightpath capacity—can be represented as one auxiliary graph.

5.6.2 Route Computation

Based on the network-state auxiliary graph with appropriate edge cost, we compute two link-disjoint paths from the access-layer output port (A_O) at the source node to the access-layer input port (A_I) at the destination node. One might think that Suurballe's algorithm can find the optimal solution for a connection request. Suurballe's algorithm, however, does not apply here because an edge in the auxiliary graph may represent a lightpath spanning a sequence of physical links. For example, edge l_2 in Fig. 5.7 traverses physical links $\langle 4, 0 \rangle$ and $\langle 0, 1 \rangle$ as in Fig. 5.4(a). As a result, multiple edges in the auxiliary graph can be in the same shared-risk link group (SRLG)[5]. Therefore, Suurballe's algorithm does not apply. Besides the SRLG constraints, the grooming-port constraint introduces additional complexity, as shown in the proof in Appendix 5.A.

A straightforward heuristic is to employ a two-step approach [Zhang et al., 2003, Li et al., 2002]: first compute a minimal-cost path as working path, and then compute a minimal-cost path as backup after removing all the links which are not SRLG-disjoint to the first minimal-cost path. The advantage of the two-step approach is that a working path is of minimal cost, which may

[5] A SRLG is a set of links which share the same risk [Strand et al., 2001].

be desirable in some situations [Li et al., 2002]. A potential drawback of the two-step approach is that it may fail in a "trap" topology [Dunn et al., 1994]. For the example network in Fig. 5.8, the two-step approach cannot find two link-disjoint paths from node 0 to node 3 (even though they exist) because the graph is disconnected after the removal of the first minimum-cost path (which is $\langle 0, 1, 2, 3 \rangle$).

We introduce backtracking based on network flow to overcome the trap situation. Let S be the set of nodes reachable from the source node after removing the links which are not SRLG-disjoint to the first minimal-cost path. Let D be the complement of S. A link is referred to as a *backhaul* link if its source node is in D and its destination node is in S. For example, link $\langle 1, 2 \rangle$ in Fig. 5.8 is a backhaul link. We make the following claim.

CLAIM 3 *If the second minimal-cost path is not found and the first minimal-cost path does not traverse any backhaul link, then there is no SRLG-disjoint path pair from the source to the destination.*

The proof is straightforward from network-flow theory (maximum flow minimum cut) and not shown here. If the second minimal-cost path is not found and backhaul links exist, our backtracking-based scheme increases the cost of the backhaul links to some large value and restarts the two-step process. This way, the first minimal-cost path will avoid, if possible, these backhaul links, and the second minimal-cost path will have a chance to reach nodes in D. For example, if we increase the cost of the backhaul link $\langle 1, 2 \rangle$ to a large number, say 10^9, and recompute the first minimal-cost path, which turns out to be $\langle 0, 1, 3 \rangle$, we are able to compute a SRLG-disjoint minimal-cost path $\langle 0, 2, 3 \rangle$.

In case there are chained trap situations, in which some traps do not appear until some others are processed, we can recursively apply the procedure. We introduce a parameter k to limit the number of recursions. The parameter k can be considered as the maximum number of trap situations we want to design our algorithm for.

A formal specification of our PAC heuristic is shown in Algorithm 5.2.

5.6.3 Lightpath-Setup Strategy

In Step 3 of Algorithm 5.2, the paths l_w and l_b may traverse a sequence of fresh, i.e., new, wavelength links. Consider an example sequence of fresh wavelength links $\langle u, \ldots, i, \ldots, v \rangle$. It can be set up as one lightpath or multiple lightpaths. If the connection to be provisioned requires the entire wavelength capacity, then we simply set up the sequence $\langle u, \ldots, i, \ldots, v \rangle$ as one lightpath; otherwise, we set it up as one or multiple lightpaths according to the following strategy.

This strategy is based on the observation that a lightpath of shorter physical hop distance is more likely to be filled up than a lightpath of longer physical hop

Algorithm 5.2 PAC

Input: $G = (V, E, C, \lambda, P)$, $c = \langle s, d, B, t_h \rangle$, existing lightpath information, wavelength usage, grooming-port usage, and k (maximum number of recursions).

Output: Two SRLG-disjoint paths, or NULL if no such paths are found.

1 If B is STS-192, then:

   (a) construct a physical reachability graph G_r^p,

   (b) apply Suurballe's algorithm on G_r^p to compute two physical link-disjoint paths l_w and l_b, and

   (c) go to Step 3 if such paths are found; otherwise, return NULL.

2 Otherwise:

   (a) construct the network-state auxiliary graph G_s based on the existing lightpath information, wavelength usage, and grooming-port usage as shown in Section 5.6.1,

   (b) compute a minimal-cost path l_w on G_s from the access-layer output vertex of node s to the access-layer input vertex of node d; return NULL if l_w is not found,

   (c) disable all the links (in G_s) which are not SRLG-disjoint to l_w,

   (d) compute a minimal-cost path l_b from the access-layer output vertex of node s to the access-layer input vertex of node d; go to Step 3 if l_b is found; return NULL if l_b is not found and $k = 0$,

   (e) identify as E_b the set of backhaul links path l_w traverses, and let $E_b^{'}$ be the set of backhaul links which share the same SRLGs as any link in E_b; (clearly, $E_b \subseteq E_b^{'}$) return NULL if set $E_b^{'}$ is empty,

   (f) increase the cost of any link in $E_b^{'}$ to some large value (e.g., 10^9), and

   (g) $k \leftarrow k - 1$; enable all the links; go to Step 2b.

3 Allocate proper resources and update network state according to the paths l_w and l_b (if necessary): update the free capacity of lightpaths involved in the paths l_w and l_b; set up new lightpaths (consume new wavelengths and free grooming ports) according to the lightpath-setup strategy shown in Section 5.6.3.

4 Return the paths l_w and l_b.

distance. To balance the physical hop distance of the lightpaths based on the limited number of available grooming ports, we introduce a parameter—threshold τ ($0 \leq \tau \leq 1$)—to determine whether to set up the sequence $\langle u, \ldots, i, \ldots, v \rangle$ as one lightpath (from node u to node v) or two lightpaths (one from node u to node i and the other from node i to node v) based on the following conditions:

$$R_i^f - 1 \geq \tau \times R_i \tag{6.1}$$

$$T_i^f - 1 \geq \tau \times T_i \tag{6.2}$$

In the above conditions, R_i and T_i denote the number of grooming-drop and grooming-add ports at node i, while R_i^f and T_i^f represent the number of available grooming-drop and grooming-add ports at node i. Conditions 6.1 and 6.2 state that, if the number of available grooming ports is not lower than the threshold after using one, node i will "break" the to-be-setup lightpath from node u to node v into two lightpaths. We can then recursively apply the above procedure to the sequences $\langle u, \ldots, i \rangle$ and $\langle i, \ldots, v \rangle$. As a result, the sequence $\langle u, \ldots, i, \ldots, v \rangle$ can be set up as multiple lightpaths, depending on the grooming-port availability at the intermediate nodes.

5.6.4 Computational Complexity

The complexity of Algorithm 5.2 is $O(k \times (|V|^2 + L))$, where L is the number of existing lightpaths in the network (L is bounded by the total number of wavelength links). Specifically, the complexity of Step 1 is $O(|V|^2)$; the complexities of Steps 2a, 2b, 2c, 2d, 2e, 2f, and 2g are $O(|E| + L)$, $O(|V|^2)$, $O(|E| + L)$, $O(|V|^2)$, $O(|E| + L)$, $O(|E|)$, and $O(|E| + L)$, respectively; the complexity of Step 3 is $O(|E| + L)$; and the complexity of Step 4 is $O(1)$.

5.7 Illustrative Numerical Results

We simulate a dynamic network environment with the assumptions that the connection-arrival process is Poisson and the connection-holding time follows a negative exponential distribution. For the illustrative results shown here, the capacity of each wavelength is STS-192; the number of the connection requests follow the distribution STS-1 : STS-3c : STS-12c : STS-48c : STS-192c = 300 : 20 : 6 : 4 : 1 (which is close to the bandwidth distribution in a practical backbone network)[6]; connection requests are uniformly distributed among all node pairs; average connection-holding time is normalized to unity; the cost of any link is unity; load (in Erlang) is defined as connection-arrival rate times average holding time times a connection's average bandwidth normalized in the unit of STS-192; and our example network topology with 16 wavelengths per fiber is shown in Fig. 5.9. 100,000 connections were simulated in each experiment.

The number of grooming ports at a node is set as the number of wavelengths times its nodal degree times a scalar Δ ($0 \leq \Delta \leq 1$). The value of Δ determines the grooming capacity of all the nodes (please refer to Section 5.2). A larger Δ indicates more grooming capable nodes. $\Delta = 1$ implies that any incoming wavelength to the W-Fabric can be dropped to the G-Fabric. The parameter

[6]Similar results were observed for other practical connection distributions, e.g., STS-1 : STS-3c : STS-12c : STS-48c : STS-192c = 5000 : 1000 : 100 : 10 : 1 for a metro-area mesh network, with appropriate adjustment in the parameters of our algorithms.

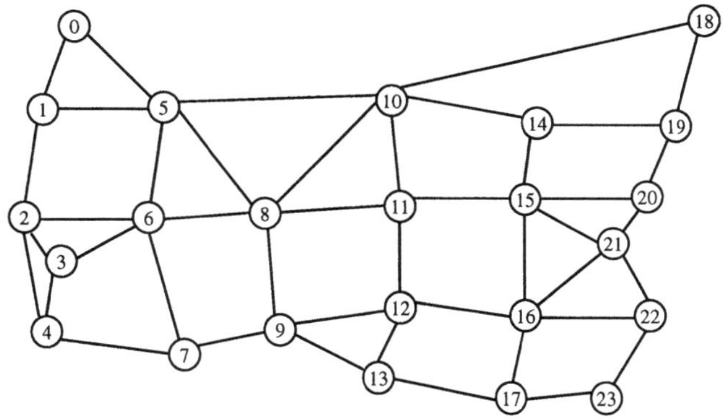

Figure 5.9. A 24-node example network topology.

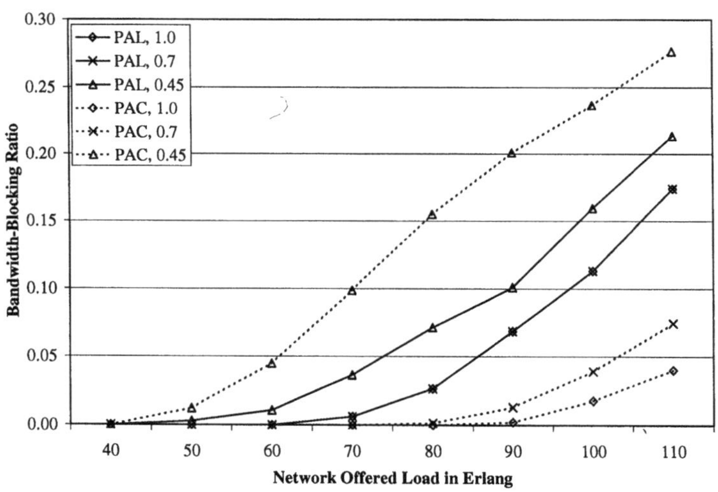

Figure 5.10. BBR for $\Delta = 1.0, 0.7$, and 0.45 (the two curves for "PAL, 1.0" and "PAL, 0.7" overlap each other).

k for PAC is set to unity since we found that the performance improvement is marginal if we increase k to any larger value. The cost-slack parameter for PAL (defined in Section 5.5.5) is $\delta = 2$. The lightpath-setup threshold (defined in Section 5.6.3) is $\tau = 1.0$. Later in Section 5.7.4, we shall examine the effect of different values of these parameters.

We now quantitatively compare PAL to PAC using the following metrics: bandwidth-blocking ratio (BBR), resource utilization, and resource-efficiency ratio (RER).

5.7.1 Bandwidth-Blocking Ratio (BBR)

BBR is defined as the amount of bandwidth blocked over the amount of bandwidth offered. Please note that pure blocking probability, defined as the percentage of the *number* of connections blocked, cannot reflect the effectiveness of the algorithm as connections have different bandwidth requirements. Figure 5.10 plots the BBR of PAL and PAC with $\Delta = 1.0, 0.7$, and 0.45. We make the following observations.

5.7.1.1 PAL vs. PAC

When the number of grooming ports is high, e.g., $\Delta = 1.0$ or 0.7, PAC has much lower BBR than PAL under moderate or high network offered load. However, when the number of grooming ports is small, e.g., $\Delta = 0.45$, PAL has much lower BBR than PAC under moderate or high network offered load. This is because PAC trades grooming ports for bandwidth efficiency in routing and grooming (please refer to Section 5.4.3).

5.7.1.2 Impact of grooming capacity on PAL

If we examine the three PAL curves in Fig. 5.10, we observe that PAL is not very sensitive to the changes in the number of grooming ports. For example, the BBRs for PAL under $\Delta = 1.0$ and $\Delta = 0.7$ are the same. When Δ further decreases to 0.45, the BBR for PAL increases moderately. The reason for this is that PAL exploits wavelengths more quickly than grooming ports. This will be further verified in Section 5.7.2.2.

5.7.1.3 Impact of grooming capacity on PAC

If we examine the three PAC curves in Fig. 5.10, we observe that PAC is very sensitive to the changes in the number of grooming ports. For example, the BBR for PAC increases moderately when Δ decreases from 1.0 to 0.7; and the BBR for PAC increases a lot when Δ further decreases to 0.45. Again, this is because PAC utilizes grooming ports more aggressively than PAL does.

5.7.2 Resource Utilization

We consider three types of resources: grooming port, wavelength, and lightpath. Grooming-port utilization is the percentage of grooming ports used. Wavelength utilization is the percentage of wavelength links utilized by lightpaths. Lightpath utilization is the percentage of a lightpath's bandwidth consumed by connections. Intuitively, lower wavelength/grooming-port utilization and higher lightpath utilization are desirable.

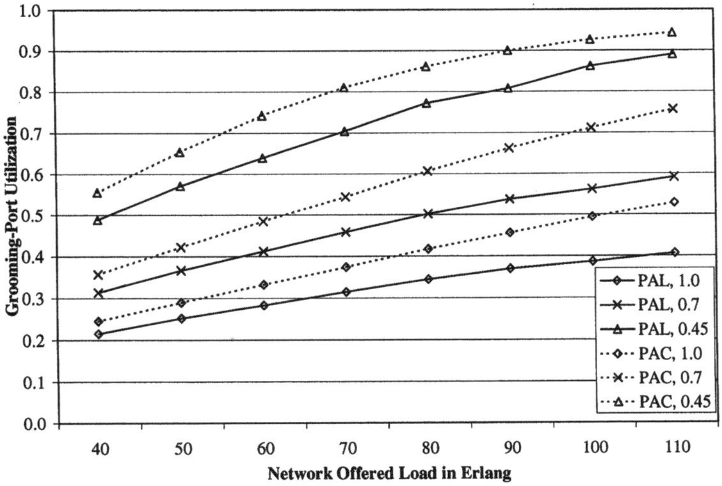

Figure 5.11. Grooming-port utilization for $\Delta = 1.0, 0.7$, and 0.45.

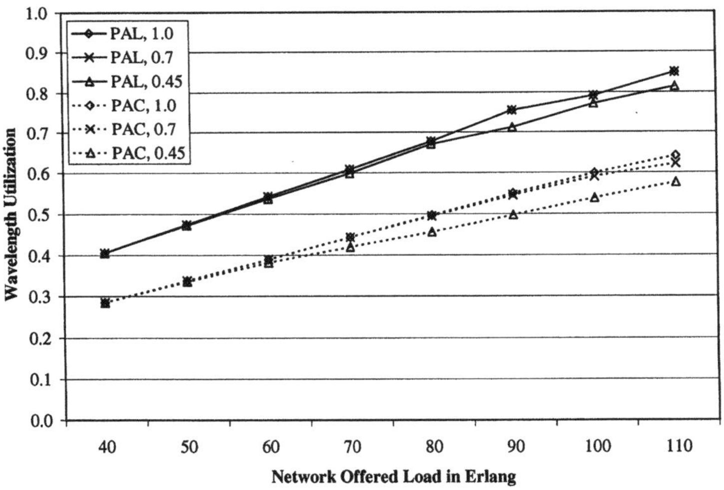

Figure 5.12. Wavelength utilization for $\Delta = 1.0, 0.7$, and 0.45.

5.7.2.1 Grooming-port utilization

Figure 5.11 plots the grooming-port utilization of PAL and PAC with $\Delta = 1.0, 0.7$, and 0.45.

If we compare PAL to PAC under the same Δ, PAC has higher grooming-port utilization. This is expected because PAC trades grooming ports for bandwidth efficiency in routing (Section 5.4.3.1). Additional results (which are not included here) indicate that a connection routed under PAC has longer av-

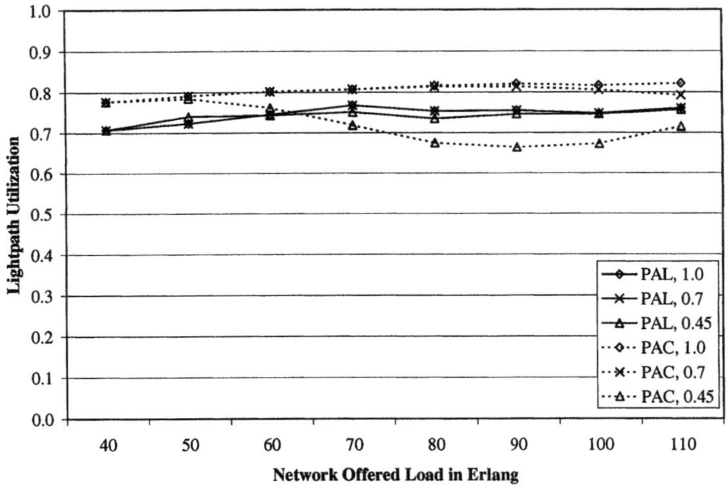

Figure 5.13. Lightpath utilization for $\Delta = 1.0, 0.7$, and 0.45.

erage lightpath-hop distance, and longer lightpath-hop distance means more grooming-port consumption.

If we examine the three curves of either PAL or PAC in Fig. 5.11, we observe that the grooming-port utilization decreases when the number of grooming ports increases. When the number of grooming ports increases to the upper bound ($\Delta = 1.0$), the grooming-port utilization under both PAL and PAC barely reaches 0.55. This implies that providing STS-1 level grooming to all the connections may not be always necessary, even when the number of lower-speed connections significantly outnumber higher-speed connections (as in our case).

5.7.2.2 Wavelength utilization

Figure 5.12 plots the wavelength utilization of PAL and PAC with $\Delta = 1.0, 0.7$, and 0.45. Our first observation is that, under the same Δ, PAL has higher wavelength utilization than PAC. This is because the additional routing constraint that a connection should be routed via a sequence of p-lightpaths under PAL incurs routing inefficiency. Our second observation is that PAL has similar wavelength utilization under different values of Δ. This is because PAL is not very sensitive to the changes in the number of grooming ports (Section 5.7.1.2). On the other hand, PAC has lower wavelength utilization when $\Delta = 0.45$ because not all the wavelengths can be exploited when the number of grooming ports is small.

5.7.2.3 Lightpath utilization

Figure 5.13 plots the lightpath utilization of PAL and PAC with $\Delta = 1.0, 0.7$, and 0.45. Under $\Delta = 1.0$ and 0.7, PAC has higher lightpath utilization because PAC is more bandwidth efficient in routing and grooming by utilizing more grooming ports. Even when the number of grooming ports is small (e.g., $\Delta = 0.45$), PAC can still have higher lightpath utilization as long as the network offered load is low (e.g., 40 Erlangs) due to the same reason. When the number of grooming ports is small and the network offered load is moderate or high, PAC quickly consumes most of the grooming ports at early stage, and connections arriving later are less likely to be groomed. Therefore, PAC has lower lightpath utilization (and much higher BBR as shown in Fig. 5.10) under such a scenario.

5.7.3 Resource-Efficiency Ratio (RER)

5.7.3.1 Definition

To better evaluate the performance of our route-computation heuristics, we introduce a new metric, called *resource-efficiency ratio* (RER) $\mathcal{E}$, which is defined as the carried load (weighted by time and normalized to STS-192 capacity) divided by the amount of allocated resources in terms of wavelength channels and grooming ports (weighted by time). This metric is defined as follows:

$$\mathcal{E}(W_\lambda, W_g) = \frac{\sum_i \rho_i \times t_i}{W_\lambda \times \sum_i \beta_i \times t_i + W_g \times \sum_i \gamma_i \times t_i}$$

where t_i is the time period between the ith event (connection arrival or departure) and $(i+1)^{th}$ event; ρ_i is the network carried load during the time period t_i; β_i is the number of wavelength links used during t_i; γ_i is the number of grooming ports used during t_i; W_λ and W_g are the relative weights of a wavelength link versus a grooming port. (Please note that ρ_i, β_i, and γ_i do not change during time period t_i as there is no other event during the period.)

Basically, $\mathcal{E}$ measures how efficiently resources have been used. If we consider "minimal hops" as our objective for a route-computation algorithm and if we assume that connections do not have any protection requirement, then the inverse of the average hop distance plus two (a connection consumes a grooming-add port and a grooming-drop port) is the upper bound for $\mathcal{E}(1,1)$. This upper bound is achieved only when every connection requires STS-192 bandwidth and follows the shortest path. Since there are limited resources (as in our case), not every connection can follow the shortest path, and the upper bound may not be achievable. If every connection requires dedicated protection, the upper bound is achieved only when every connection requires STS-192 and follows the shortest link-disjoint paths. For the topology in Fig. 5.9, the upper bound for $\mathcal{E}(1,1)$ with dedicated protection is $\frac{1}{11}$.

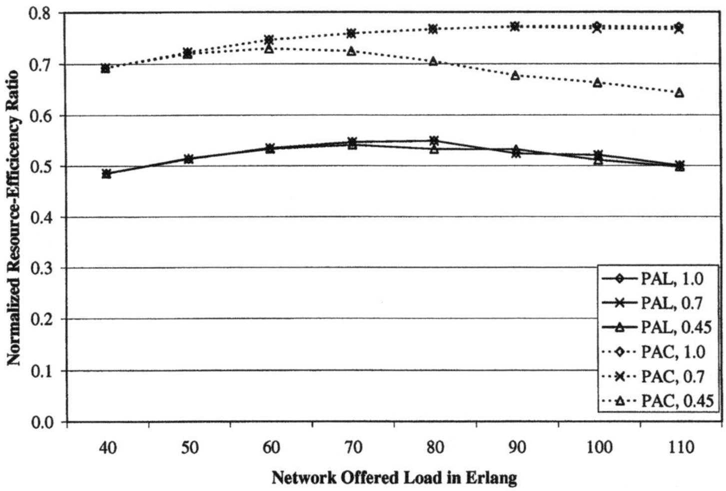

Figure 5.14. Normalized RER $\mathcal{E}(1,0)$ for $\Delta = 1.0, 0.7$, and 0.45.

5.7.3.2 Wavelength efficiency

If $W_\lambda = 1$ and $W_g = 0$, RER $\mathcal{E}(1,0)$ measures how efficiently wavelength channels have been utilized. RER $\mathcal{E}(1,0)$ is a more comprehensive metric compared to wavelength utilization and lightpath utilization. Figure 5.14 plots the normalized RER $\mathcal{E}(1,0)$ for $\Delta = 1.0, 0.7$, and 0.45. Under the same Δ, PAL has lower RER $\mathcal{E}(1,0)$ than PAC. This is because PAL has the additional routing constraint that the survivable route for a connection should be a concatenation of p-lightpaths.

5.7.3.3 Grooming-port efficiency

If $W_\lambda = 0$ and $W_g = 1$, RER $\mathcal{E}(0,1)$ measures how efficiently grooming ports have been utilized. Figure 5.15 plots the normalized RER $\mathcal{E}(0,1)$ for $\Delta = 1.0, 0.7$, and 0.45. The observation that PAC has lower RER $\mathcal{E}(0,1)$ under the same Δ verifies our results in Section 5.7.2.1 that PAC requires more grooming ports.

5.7.3.4 Tradeoff between wavelengths and grooming ports

As stated earlier in Section 5.4, PAL and PAC trade off between wavelengths and grooming ports. Below, we show that either PAL and PAC can have higher RER, depending on the relative weights of a wavelength channel (W_λ) and a grooming port (W_g). Figures 5.16 and 5.17 plot the normalized RER $\mathcal{E}(2,1)$ and $\mathcal{E}(1,4)$ for $\Delta = 1.0, 0.7$, and, 0.45, respectively. We observe that (1) PAC has higher RER when a wavelength channel weighs more than a grooming port (e.g., $W_\lambda : W_g = 2 : 1$), and (2) PAL has higher RER when a grooming

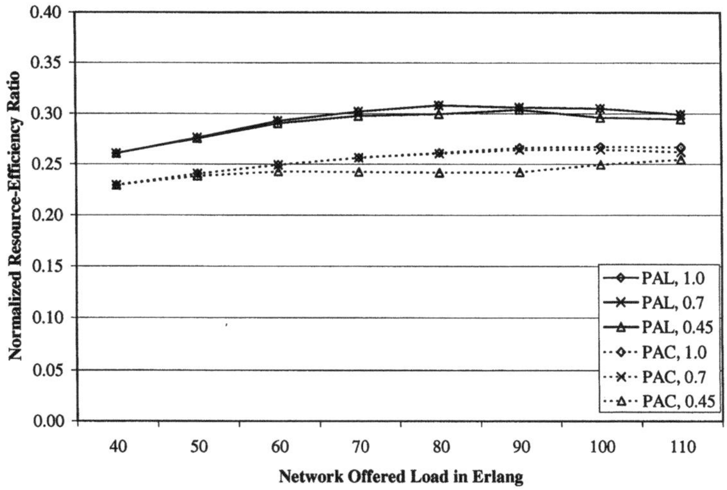

Figure 5.15. Normalized RER $\mathcal{E}(0,1)$ for $\Delta = 1.0, 0.7$, and 0.45.

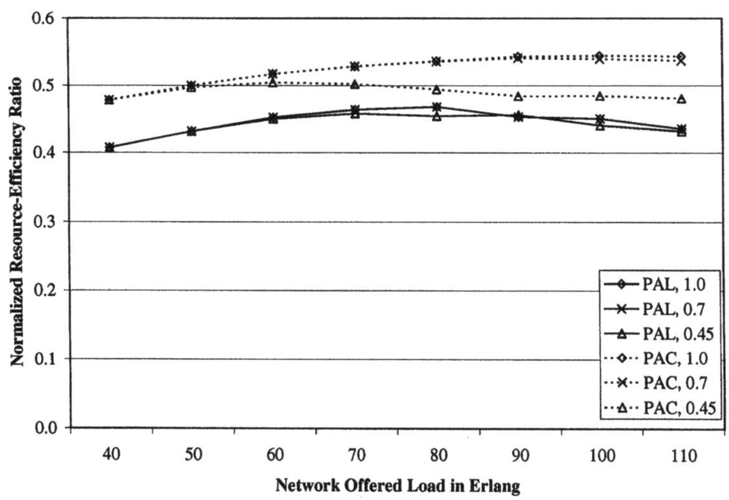

Figure 5.16. Normalized RER $\mathcal{E}(2,1)$ for $\Delta = 1.0, 0.7$, and 0.45.

port weighs a lot more than a wavelength channel (e.g., $.W_\lambda : W_g = 1 : 4$). These results confirm our analysis in Section 5.4.3 that PAL trades bandwidth efficiency in routing (or utilizing wavelengths) for the savings in grooming port and PAC trades grooming ports for bandwidth efficiency in routing and grooming.

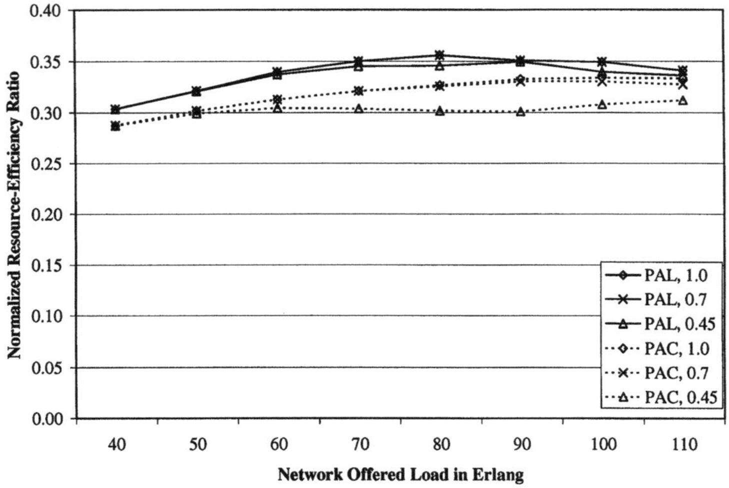

Figure 5.17. Normalized RER $\mathcal{E}(1,4)$ for $\Delta = 1.0, 0.7$, and 0.45.

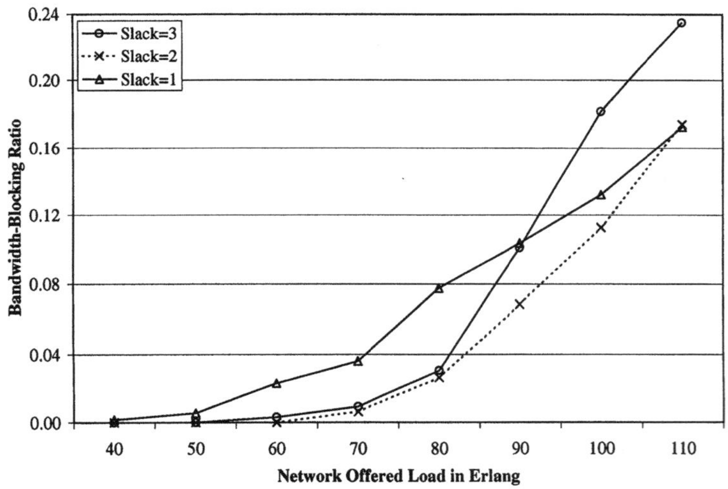

Figure 5.18. BBR of PAL for $\delta = 3, 2$, and 1 ($\Delta = 1.0$).

5.7.4 Effect of Different Parameters

Further performance improvements are possible by fine tuning the parameters in Algorithms 5.1 and 5.2.

5.7.4.1 Cost-slack parameter δ in PAL

Figure 5.18 plots the BBR of PAL with $\Delta = 1.0$ under different values of the cost-slack parameter δ. Similar results have been observed for other values

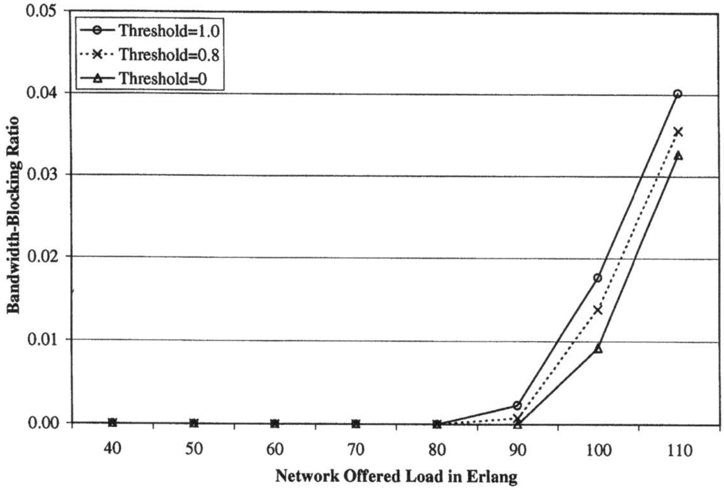

Figure 5.19. BBR of PAC for $\tau = 1.0, 0.8$, and 0 ($\Delta = 1.0$).

of Δ. Recall that the cost-slack parameter represents the preference of using an existing p-lightpath over setting up a new p-lightpath (Section 5.5.5).

Our first observation is that the BBR is much higher with $\delta = 1$ under low or moderate network offered load (e.g., 40-90 Erlangs). When $\delta = 1$, Algorithm 5.1 exploits the free wavelengths quickly, and many lightpaths are only lightly loaded. Accordingly, high-speed connections (e.g., STS-192 connections) are more likely to be blocked. Additional results (not shown here) indicate that the lightpath utilization is much lower, and the wavelength utilization is much higher (between the load region 40-90 Erlangs) when $\delta = 1$.

Our second observation is that BBR increases much faster for larger cost-slack parameters. When the cost-slack parameter δ increases, Algorithm 5.1 prefers to use an existing p-lightpath over setting up a new p-lightpath even though the existing p-lightpath costs δ more than the new one. Consequently, while the existing p-lightpaths are filled up more perfectly, the route for a connection under larger δ is less optimal than the route under smaller δ. When the network offered load is high (e.g., over 90 Erlangs), the inefficiency in routing dominates and the BBR is higher for larger cost-slack parameters.

5.7.4.2 Threshold τ in PAC

The performance of PAC can be improved by properly adjusting the threshold τ (Section 5.6.3) when setting up lightpaths. Figure 5.19 plots the BBR of PAC with $\Delta = 1.0$ under different values of the threshold τ. We observe that the BBR decreases when the threshold τ reduces from 1.0 to 0.8 to 0. When the threshold τ decreases, the average physical hop distance of lightpaths reduces since there are sufficient number of grooming ports ($\Delta = 1.0$). As a result, more low-speed connections are likely to be grooming onto existing lightpaths and the BBR decreases. More results (not shown here) indicate that lower threshold τ leads to higher lightpath utilization and lower wavelength utilization when $\Delta = 1.0$. Interestingly, the grooming-port utilization under $\tau = 0.8$ is lower than the grooming-port utilization under $\tau = 1.0$. This is due to the difference between global optimization (as is the case when $\tau = 0.8$) and local optimization (as is the case when $\tau = 1.0$): setting up lightpaths with short physical hop distance initially ($\tau = 0.8$) potentially benefits later connections, and less number of lightpaths are needed because existing lightpaths can be filled up more perfectly. If the threshold is very low, e.g., $\tau = 0$, then many grooming ports need to be used and the grooming-port utilization can be high.

When the number of grooming ports is not sufficient, e.g., $\Delta = 0.7$, similar results have been observed for threshold τ close to unity. In general, the less the number of grooming ports, the closer to unity the threshold should be.

5.8 Conclusion

We investigated the problem of survivable traffic grooming for optical WDM mesh networks with dedicated protection. We proposed two approaches—PAL and PAC—for grooming a connection request with dedicated protection based on a generic grooming-node architecture. We proved that the problem of provisioning a connection under PAC is $\mathcal{NP}$-complete, and developed effective heuristics for both schemes. Comparisons between PAL and PAC uncovered the following findings. Under today's typical connection-bandwidth distribution where lower-bandwidth connections outnumber higher-bandwidth connections, PAC outperforms PAL (in terms of bandwidth-blocking ratio, lightpath utilization, and wavelength utilization) if the number of grooming ports is large; however, PAL outperforms PAC (in terms of bandwidth-blocking ratio and grooming-port utilization) when the number of grooming ports is moderate or small.

In the next chapter, we investigate the problem of survivable traffic grooming with shared protection.

APPENDIX 5.A: NP-Completeness of WDM-PAC

In WDM-PAC, we are given a virtual (lightpath) topology, a physical (available wavelength) topology, grooming ports at each node, a source node s, and a destination node d. If we view the virtual topology and the physical topology as two layers of the same topology, the WDM-PAC problem is to decide whether there exist two link-disjoint paths from node s to node d on this two-layered (virtual and physical) graph. We reduce the known $\mathcal{NP}$-complete problem, the WDM-DISJOINT-PATH problem [Zhang et al., 2003], to WDM-PAC. The existence version of the WDM-DISJOINT-PATH problem is as follows:

Instance:

1 A WDM mesh network, $G = (V, E)$, where V is the set of WDM network nodes, E is the set of fiber links, and each fiber link in the network has $W = 2$ wavelengths[7].

2 The routes and the wavelengths of the existing lightpaths in the network.

3 A source node s and a destination node d.

Question: Does there exist a pair of link-disjoint, wavelength-continuous lightpaths from node s to node d?

Proof: WDM-PAC $\in \mathcal{NP}$ since a non-deterministic algorithm can guess two paths between node s and node d and check in polynomial time if the two paths are physically link-disjoint and obey the resource constraints (wavelengths and grooming ports).

Given an arbitrary instance of WDM-DISJOINT-PATH $G = (V, E), W = 2, s, d$, and the routes and the wavelengths of the existing lightpaths, construct an instance of WDM-PAC as follows: The virtual topology is one layer of G and the physical topology is the other layer of G (we can consider the given graph G as a two-layered—λ_1 and λ_2—graph since it has two wavelengths); the number of free grooming-add ports at node s is two and the number of free grooming-drop ports at node d is two; the number of free grooming ports at other nodes is zero; and every link on the virtual topology has sufficient amount of free capacity.

If there exist two link-disjoint paths in the instance of WDM-DISJOINT-PATH, then the two paths are also link-disjoint in the instance of WDM-PAC and they obey wavelength-availability constraints. These two paths obey grooming-port constraints because: (1) in either case, node s needs two grooming-add ports and node d needs two grooming-drop ports; (2) due to the wavelength-continuity constraint, each of the two paths must stay on the same wavelength, i.e., they stay on the same layered graph. This implies that there is no need for grooming ports to originate/terminate the two paths except at the source and destination nodes. Hence, the two paths form a valid solution for the WDM-PAC instance.

If there exist two link-disjoint paths in the instance of WDM-PAC, then the two paths are also link-disjoint in the instance of WDM-DISJOINT-PATH and they respect the wavelength-availability constraint. Since only the source and the destination nodes have free grooming ports, none of the two paths can utilize links from both layers. Thus, the two paths are wavelength continuous. Hence, the two paths are a valid solution for the WDM-DISJOINT-PATH instance.

This concludes our proof that WDM-PAC is $\mathcal{NP}$-complete. ■

[7]The WDM-DISJOINT-PATH problem is $\mathcal{NP}$-complete for any $W > 1$. We use $W = 2$ here for convenience.

Chapter 6

SURVIVABLE TRAFFIC GROOMING–SHARED PROTECTION

The previous chapter explored survivable traffic grooming with dedicated protection. This chapter investigates survivable traffic grooming with shared protection.

We propose three approaches—protection-at-lightpath (PAL) level, mixed protection-at-connection (MPAC) level, and separate protection-at-connection (SPAC) level—for dynamically provisioning shared-protected subwavelength-granularity connection requests against single-fiber failures. We investigate their characteristics under the same grooming-node architecture in Fig. 5.1 and design efficient heuristics. Our work differs from previous work in that we focus on route computation, the impact of different backup-sharing approaches, and the tradeoff between wavelength and grooming capacity.

The rest of this chapter is organized as follows. Section 6.1 formally states the problem. Section 6.2 presents three approaches—protection-at-lightpath (PAL) level, mixed protection-at-connection (MPAC) level, and separate protection-at-connection (SPAC) level—and provides a qualitative comparison. Section 6.3 presents heuristic algorithms for PAL, MPAC, and SPAC. Section 6.4 compares the three schemes under different network configurations. Section 6.5 concludes this chapter.

6.1 Problem Statement

We first define the notations and then formally state the dynamic connection-provisioning problem. A network is represented as a weighted, directed graph $G = (V, E, C, \lambda, P)$, where V is the set of nodes, E is the set of unidirectional fibers (referred to as links), $C : E \rightarrow R^+$ is the cost function for each link (where R^+ denotes the set of positive real numbers), $\lambda : E \rightarrow Z^+$ specifies the number of wavelengths on each link (where Z^+ denotes the set of positive integers), and $P : V \rightarrow Z^+$ specifies the number of grooming ports at each node.

A connection request is represented as a quadruple $\langle s, d, B, t_h \rangle$, which specifies the source node, the destination node, the bandwidth requirement, and the holding time in this order. In this study, every connection needs to be protected, and the backup resources can be shared.

We now formally state the dynamic connection-provisioning problem as follows: Given the current network state (which includes the network topology as a weighted digraph G, existing lightpath/connection information [e.g., routes and wavelengths, etc.], wavelength usage, and grooming-port usage), route each new connection request with respect to its bandwidth and protection requirements (shared protection) while minimizing the incremental cost in terms of the total cost of the working and backup paths under the assumptions that existing connections cannot be disturbed and information about future arrivals is not known.

The existence version of the above problem for provisioning one connection request under the current network state is $\mathcal{NP}$-complete. This is because a special case of this problem, in which the number of grooming ports is sufficient and every connection request requires full wavelength capacity, has been proven to be $\mathcal{NP}$-complete in [Ou et al., 2003]. Thus, practical heuristics are needed.

6.2 Proposed Schemes

To provision a connection request, there are two types of resource constraints—wavelengths and grooming ports. Typically, the more the number of wavelengths the network has, the less the number of grooming ports a node needs, and vice versa.

We propose three schemes—protection-at-lightpath (PAL) level, mixed protection-at-connection (MPAC) level, and separate protection-at-connection level (SPAC)—for provisioning shared-protected connection requests. These three schemes explore different ways of backup routing and the tradeoff between wavelengths and grooming ports.

Below, we illustrate the three schemes via an example. For the initial network configuration shown in Fig. 6.1, every edge corresponds to a bidirectional fiber; each fiber has two wavelengths; the wavelength capacity is STS-192; every node has three grooming ports (T and R represent the number of available grooming-add and grooming-drop ports, respectively).

6.2.1 Protection-at-Lightpath (PAL) Level

6.2.1.1 Basic idea

PAL provides end-to-end protection with respect to lightpath. Under PAL, a connection is routed through a sequence of protected lightpaths, or p-lightpaths.

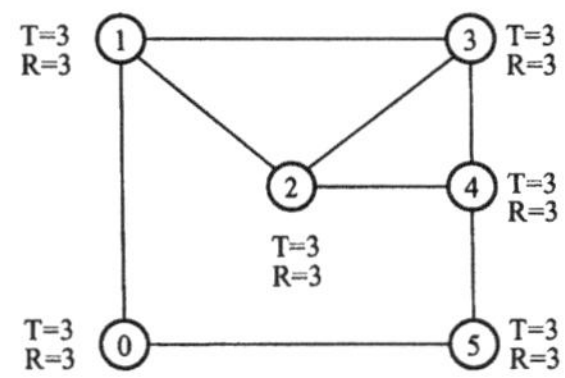

Figure 6.1. Example: initial network configuration.

A p-lightpath has a *lightpath* as working path and a link-disjoint *path* as backup path. For example, in Fig. 6.2(a), p-lightpath l_1 has lightpath $\langle 0,1,2\rangle$ as working path and path $\langle 0,5,4,2\rangle$ as backup path. Please note the differences between the working path and the backup path of a p-lightpath. The working path of a p-lightpath is set up as a lightpath during normal operation. Therefore, as a lightpath does, the working path consumes a grooming-add port at the source node and a grooming-drop port at the destination node of a p-lightpath; and the working path of a p-lightpath bypasses any intermediate nodes along its path. However, the backup path of a p-lightpath is not set up as a lightpath during normal operation. Therefore, the backup path of a p-lightpath does not consume any grooming port; and wavelengths along a backup path are only reserved. In case the working path fails, protection switching occurs at lightpath level and the backup path is set up as a lightpath by utilizing the grooming ports previously used by the working path.

Two p-lightpaths can share wavelengths along common backup links if their working paths are link-disjoint. Clearly, a connection routed under PAL survives from single-link failures since each p-lightpath survives from single-link failures by definition. Since protection occurs at lightpath level, PAL has the advantages of low implementation complexity and low signaling overhead when a failure occurs. This will be further elaborated in Section 6.2.4. Below, we illustrate PAL in more detail by provisioning three connection requests.

6.2.1.2 Example

Upon the arrival of the first connection request c_1, $\langle 0,2,\text{STS-12c},t_1\rangle$, one way of provisioning c_1 under PAL is shown in Fig. 6.2(a). Connection c_1 is routed via p-lightpath l_1, which has lightpath $\langle 0,1,2\rangle$ as working path and path $\langle 0,5,4,2\rangle$ as backup path. p-lightpath l_1 consumes a grooming-add port at node 0 and a grooming-drop node at node 2. The free capacity of p-lightpath l_1 is STS-180.

Suppose that c_1 remains in the network when the second connection request c_2, $\langle 0,3,\text{STS-3c},t_2\rangle$, arrives. One way of provisioning c_2 under the current network state is shown in Fig. 6.2(b). Connection c_2 is routed via p-lightpath l_2, which has lightpath $\langle 0,1,3\rangle$ as working path and path $\langle 0,5,4,3\rangle$ as backup path.

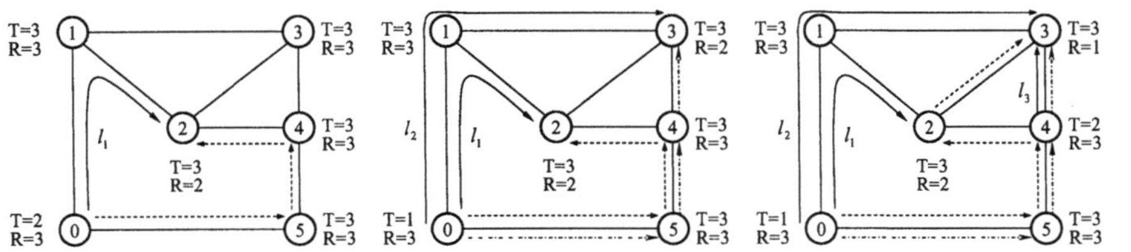

(a) after provisioning c_1 (b) after provisioning c_2 (c) after provisioning c_3

Figure 6.2. PAL: provisioning connections $c_1(\langle 0, 2, \text{STS-12c}, t_1\rangle)$, $c_2(\langle 0, 3, \text{STS-3c}, t_2\rangle)$, and $c_3(\langle 4, 3, \text{STS-48c}, t_3\rangle)$.

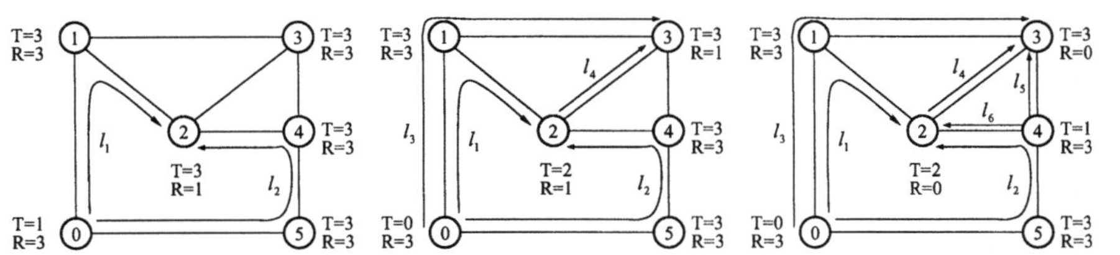

(a) after provisioning c_1 (b) after provisioning c_2 (c) after provisioning c_3

Figure 6.3. MPAC: provisioning connections $c_1(\langle 0, 2, \text{STS-12c}, t_1\rangle)$, $c_2(\langle 0, 3, \text{STS-3c}, t_2\rangle)$, and $c_3(\langle 4, 3, \text{STS-48c}, t_3\rangle)$.

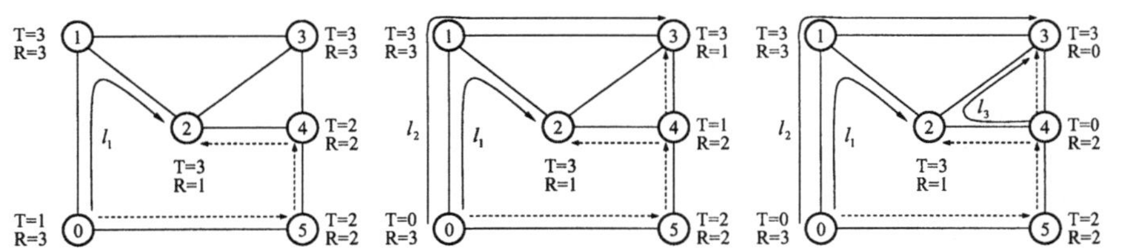

(a) after provisioning c_1 (b) after provisioning c_2 (c) after provisioning c_3

Figure 6.4. SPAC: provisioning connections $c_1(\langle 0, 2, \text{STS-12c}, t_1\rangle)$, $c_2(\langle 0, 3, \text{STS-3c}, t_2\rangle)$, and $c_3(\langle 4, 3, \text{STS-48c}, t_3\rangle)$.

p-lightpath l_2 consumes a grooming-add port at node 0 and a grooming-drop port at node 3. The free capacity of p-lightpath l_2 is STS-189. Two wavelengths need to be reserved along links $\langle 0, 5\rangle$ and $\langle 5, 4\rangle$ because (1) the working paths of p-lightpaths l_1 and l_2 traverse common link $\langle 0, 1\rangle$, and (2) protection occurs at lightpath level, i.e., backup sharing only occurs at wavelength level.

Suppose that c_1 and c_2 remain in the network when the third connection request c_3, $\langle 4, 3, \text{STS-48c}, t_3\rangle$, arrives. One way of provisioning c_3 under the current network state is shown in Fig. 6.2(c). Connection c_3 is routed via p-lightpath l_3, which has lightpath $\langle 4, 3\rangle$ as working path and path $\langle 4, 2, 3\rangle$ as backup path. p-lightpath l_3 consumes a grooming-add port at node 4 and a grooming-drop port at node 3. The free capacity of p-lightpath l_3 is STS-144.

Please note that the backup paths of p-lightpaths l_1 and l_3 share the wavelength reserved along link $\langle 4, 2 \rangle$.

6.2.2 Mixed Protection-at-Connection (MPAC) Level

6.2.2.1 Basic idea

MPAC and SPAC provide end-to-end protection with respect to connection. Under MPAC, a connection is routed via link-disjoint working and backup paths, each of which traverses a sequence of lightpaths. A lightpath traversed by a working path utilizes a portion of its capacity to carry traffic for that working path during normal operation. A lightpath traversed by a backup path reserves part of its capacity for that backup path. The backup capacity a lightpath reserves can be shared among multiple backup paths provided that their corresponding working paths are link-disjoint. In this context, "mixed" means that the capacity of one wavelength can be utilized by *both* working paths and backup paths; "separate", on the other hand, means that the capacity of a wavelength can be utilized by *either* working paths or backup paths, but not both. MPAC seems to be the most intuitive approach since it deals with individual connections and therefore can pack connections efficiently. Later in Section 6.2.4, we shall show that it may not achieve the best performance due to the intricacy of backup sharing. Below, we illustrate MPAC in more detail using the same example as before.

6.2.2.2 Example

When the first connection request $c_1, \langle 0, 2, \text{STS-12c}, t_1 \rangle$ arrives, one way of provisioning c_1 under MPAC is shown in Fig. 6.3(a). The working and backup paths of connection c_1 traverse lightpaths l_1 and l_2, respectively. The free capacity of both lightpaths l_1 and l_2 is STS-180. The backup capacity reserved on lightpath l_1 is zero. The backup capacity reserved on lightpath l_2 is STS-12c, and it is used to protect connection c_1's working path. Both lightpaths l_1 and l_2 consume a grooming-add port at node 0 and a grooming-drop port at node 2.

Suppose that connection c_1 remains in the network when connection request $c_2, \langle 0, 3, \text{STS-3c}, t_2 \rangle$, arrives. One possible solution of provisioning c_2 under MPAC is shown in Fig. 6.3(b). Connection c_2 is routed via two link-disjoint paths—lightpath l_3 and the two-lightpath sequence $\langle l_2, l_4 \rangle$. The working path can traverse either of the two paths, say lightpath l_3. The free capacity of both lightpaths l_3 and l_4 is STS-189. The free capacity of lightpath l_2 is STS-177 since lightpaths l_1 and l_3 traverse common link $\langle 0, 1 \rangle$. The backup capacity on lightpath l_2 is STS-15 (STS-12c capacity is used to protect the working path of connection c_1, and STS-3c capacity is used to protect the working path of c_2). The backup capacity on lightpath l_4 is STS-3c and it is used to protect the working path of connection c_2.

Suppose that connections c_1 and c_2 remain in the network when connection request c_3, $\langle 4, 3, \text{STS-48c}, t_3 \rangle$, arrives. One way of provisioning c_3 under MPAC is shown in Fig. 6.3(c). The working path of c_3 traverses lightpath l_5 and the backup path traverses the two-lightpath sequence $\langle l_6, l_4 \rangle$. The free capacity of lightpaths l_5 and l_6 is STS-144. The free capacity of lightpath l_4 is STS-144 because the backup paths of connections c_2 and c_3 can share backup capacity (c_2's working path, l_3, and c_3's working path, l_5, are link-disjoint). The backup capacity of lightpath l_4 is STS-48 (STS-48c capacity is used to protect the working path of connection c_3, and STS-3c capacity—shared with the backup path of c_3—is used to protect the working path of connection c_2). The backup capacity on lightpath l_6 is STS-48c and it is used to protect c_3's working path.

6.2.3 Separate Protection-at-Connection (SPAC) Level

6.2.3.1 Basic idea

SPAC provides end-to-end protection with respect to connection. Under SPAC, a connection is routed via link-disjoint working and backup paths. A working path traverses a sequence of lightpaths. A backup path traverses a sequence of links, each of which has judiciously reserved a number of wavelengths as backup resources. (This differs from MPAC, in which a backup path traverses a sequence of lightpaths.) In addition, a grooming-add port at the source end of the link and a grooming-drop port at the destination end of the link need to be reserved for each reserved wavelength because multiple backup paths groomed onto the same wavelength on a link may go to different next hops. In this context, "separate" means that the capacity of a wavelength can be utilized by *either* working paths or backup paths, but not both. SPAC is deliberately constructed in a way to trade grooming ports for increased backup sharing, as will be elaborated in Section 6.2.4. Below, we illustrate SPAC in more detail using the same example as before.

6.2.3.2 Example

When the first connection request c_1, $\langle 0, 2, \text{STS-12c}, t_1 \rangle$ arrives, one way of provisioning c_1 under SPAC is shown in Fig. 6.4(a). The working path of connection c_1 traverses lightpath l_1, and the backup path traverses path $\langle 0, 5, 4, 2 \rangle$. The free capacity of lightpath l_1 is STS-180. Every link along the backup path needs to reserve one wavelength as backup capacity, while only STS-12c of the entire wavelength capacity is used to protect c_1's working path. For every link along the backup path, the upstream node needs to reserve one grooming-add port and the downstream node needs to reserve one grooming-drop port since one more wavelength has been reserved.

Suppose that connection c_1 remains in the network when connection request $c_2, \langle 0, 3, \text{STS-3c}, t_2 \rangle$, arrives. One possible solution of provisioning c_2 under SPAC is shown in Fig. 6.4(b). The working path of connection c_2 traverses lightpath l_2, and the backup path traverses path $\langle 0, 5, 4, 3 \rangle$. The free capacity of lightpath l_2 is STS-189. One wavelength along link $\langle 4, 3 \rangle$ needs to be reserved as backup capacity, STS-3c capacity of which is used to protect c_2's working path. STS-15 capacity of the entire backup capacity along links $\langle 0, 5 \rangle$ and $\langle 5, 4 \rangle$ is used to protect the working paths of connections c_1 and c_2 (STS-12c for c_1 and STS-3c for c_2).

This step demonstrates why one grooming-drop port and two grooming-add ports need to be reserved at node 4. If link $\langle 0, 1 \rangle$ fails, connection c_1 needs to be rerouted along $\langle 0, 5, 4, 2 \rangle$, and connection c_2 needs to be rerouted along $\langle 0, 5, 4, 3 \rangle$. As a result, node 4 needs to drop one wavelength, say λ_1, to the G-Fabric via one grooming-drop port. After unpacking wavelength λ_1, the G-Fabric grooms connection c_1 to an appropriate wavelength, say λ_2, and adds wavelength λ_2 to the W-Fabric via one grooming-add port; the G-Fabric also grooms connection c_2 to an appropriate wavelength, say λ_3, and inserts wavelength λ_3 to the W-Fabric via another grooming-add port. Then, the W-Fabric switches wavelength λ_2 to the outgoing port towards node 2 and wavelength λ_3 to the outgoing port towards node 3. In general, one grooming-add port and one grooming-drop port are needed for each reserved wavelengths on a link.

Suppose that connections c_1 and c_2 remain in the network when connection request $c_3, \langle 4, 3, \text{STS-48c}, t_3 \rangle$, arrives. One way of provisioning c_3 under SPAC is shown in Fig. 6.4(c). The working path of connection c_3 traverses lightpath l_3 and the backup path traverses path $\langle 4, 3 \rangle$. The free capacity of lightpath l_3 is STS-144. STS-48c of the entire backup capacity along link $\langle 4, 3 \rangle$ is used to protect the working path of c_3; out of this STS-48c capacity, STS-3c is also used to protect the working path of c_2.

6.2.4 A Qualitative Comparison

The above illustrative examples indicate that the three schemes perform differently in terms of routing and the amount of resources required. Below, we qualitatively compare their characteristics with respect to routing, backup sharing, and operational complexity. For convenience, we will use the term "PAC" to refer to both MPAC and SPAC hereafter, whenever appropriate, because they have several similar properties.

6.2.4.1 Routing

The difference in routing between PAL and PAC is that PAL provides end-to-end protection with respect to lightpath while PAC provides end-to-end protection with respect to connection. Under PAL, when a failure occurs, the end nodes of the affected p-lightpaths first configure their backup paths and then switch over; the affected connections are oblivious to the protection-switching process. Under PAC, when a failure occurs, the end-nodes of the affected connections (which could be significantly more than the number of affected p-lightpaths) first configure their backup paths and then switch over.

Please note that a connection from node s to node d routed under PAL may not have two link-disjoint paths between node s and node d, while the connection still survives from single-link failures. For example, suppose that the connection is routed via two p-lightpaths l_1 and l_2 under PAL. The concatenation of the working paths of p-lightpaths l_1 and l_2 may not be link-disjoint from the concatenation of the backup paths of p-lightpaths l_1 and l_2. This is because the working path of l_1 and the backup path of l_2 (or the working path of l_2 and the backup path of l_1) can traverse common links.

Routing-wise, PAL performs at an aggregate level (lightpath) and PAC performs on a per-flow (connection) basis. As a result, PAL trades the bandwidth efficiency in routing each specific sub-wavelength connection request for the savings in grooming-port usage. In PAL, the backup path of a p-lightpath does not require any grooming port. When a fiber along the working path of a p-lightpath fails, all of the traffic carried by the failed working path can be rerouted to the backup path of that p-lightpath, and the grooming ports (at the end nodes of the p-lightpath) previously used by the working path can be reused by the backup path. However, in SPAC, the end nodes of a link need to reserve a grooming-add/drop port for each reserved wavelength because multiple backup paths groomed onto the same wavelength on a link may go to different next hops; in MPAC, each lightpath reserves a portion of its bandwidth as backup capacity, thus backup capacity consumes a fraction of the grooming ports.

6.2.4.2 Backup sharing

MPAC differs from PAL and SPAC in backup sharing. The backup path of a connection under MPAC is the concatenation of lightpaths. The backup path of a connection under SPAC (or the backup path of a p-lightpath under PAL) is the concatenation of links with reserved wavelengths. This difference in backup routing has two implications on backup sharing. First, since a lightpath may span multiple links, the backup capacity reserved on a lightpath (as in MPAC) is less likely to be shared among multiple connections than the backup capacity reserved on a link (as in PAL and SPAC).

The second implication applies to wavelength-convertible networks only. Under MPAC, the backup path of a connection traverses a sequence of lightpaths, thus a backup path has both fixed route and fixed wavelength assignment [Doverspike and Yates, 2001]. Under SPAC, the backup path of a connection (or the backup path of a p-lightpath under PAL) traverses a sequence of links with a number of reserved wavelengths, thus a backup path has only fixed route but not fixed wavelength assignment [Datta et al., 2001]. Basically, under

SPAC and PAL, the reserved wavelengths on a link act like a "pool" for all the failure scenarios, and backup-capacity sharing among different wavelengths on a link is facilitated by the existence of wavelength converters. However, under MPAC, the backup-capacity sharing among different wavelengths on a link is not possible because backup capacity resides inside lightpaths, and multiple lightpaths cannot share their reserved backup capacity. We illustrate this difference in the following example.

Consider the changes in backup capacity on an arbitrary link $\langle u, v \rangle$ in a hypothetical network. Suppose that STS-156 capacity will be rerouted on link $\langle u, v \rangle$ when some other link $\langle x, y \rangle$ in this network fails; STS-108 capacity will be rerouted on link $\langle u, v \rangle$ when some other link $\langle i, j \rangle$ in this network fails; and no more than STS-108 capacity will be rerouted on link $\langle u, v \rangle$ when any other link fails. Clearly, STS-156 backup capacity needs to be reserved along link $\langle u, v \rangle$, assuming that any link is not in the same shared-risk link group (SRLG)[1] as any other link. Under SPAC, link $\langle u, v \rangle$ needs to reserve one wavelength; under MPAC, a lightpath, l_1, from node u to node v needs to be set up. When a new connection request c_1, $\langle i, j, \text{STS-48c}, t_h \rangle$, arrives, suppose that its working path traverses link $\langle i, j \rangle$ and backup path traverses link $\langle u, v \rangle$ in this hypothetical network under both SPAC and MPAC. Since only STS-156 capacity needs to be rerouted when link $\langle i, j \rangle$ fails, no more backup capacity needs to be reserved under both SPAC and MPAC. Assume that connection c_1 remains in the network when another connection request c_2, $\langle i, j, \text{STS-48c}, t'_h \rangle$, arrives. Suppose that connection c_2's working path traverses link $\langle i, j \rangle$ and backup path traverses link $\langle u, v \rangle$ in this hypothetical network under both SPAC and MPAC. Then, STS-204 capacity will be rerouted on link $\langle u, v \rangle$ if link $\langle i, j \rangle$ fails. As a result, under SPAC, link $\langle u, v \rangle$ needs to reserve two wavelengths (since wavelength capacity is STS-192), which combine to provide STS-204 backup capacity. Under MPAC, another lightpath, l_2, from node u to node v needs to be set up. Lightpath l_1 reserves STS-156 capacity and lightpath l_2 reserves STS-48 capacity as backup capacity.

The difference appears when connection c_1 leaves and connection c_2 remains in the network. Only STS-156 capacity will be rerouted on link $\langle u, v \rangle$ when either link $\langle x, y \rangle$ or link $\langle i, j \rangle$ fails after connection c_1 leaves. Consequently, under SPAC, only one wavelength needs to be reserved on link $\langle u, v \rangle$, and another previously reserved wavelength can be released. Under MPAC, however, lightpath l_1 still needs to reserve STS-156 backup capacity since STS-156 capacity will be rerouted on lightpath l_1 when link $\langle x, y \rangle$ fails. Without reconfiguring connection c_2's backup path, lightpath l_2 still needs to reserve STS-48 backup capacity. As a result, under MPAC, STS-204 capacity has been reserved while

[1]A SRLG is a set of links which share the same risk [Strand et al., 2001].

only STS-156 is really needed. (PAL will perform similarly to SPAC in this example.)

In short, PAL and SPAC trade the flexibility in grooming for the freedom in backup sharing. Under PAL and SPAC, working paths are groomed onto lightpaths while backup paths are groomed onto reserved wavelengths. However, MPAC has the flexibility in grooming working paths and backup paths (of different connections) onto the same lightpath.

6.2.4.3 Operational complexity

From implementation point of view, PAL is simpler than PAC as PAL demands less information in route computation. While both PAL and PAC need the routing information of all the existing lightpaths to provision a shared-protected connection request, PAL does not require any information about the existing connections. PAC, however, does require the detailed routing information of all the existing connections. Under PAL, the routing information of the working paths of two p-lightpaths is sufficient to determine whether the backup paths of these two p-lightpaths can share wavelengths along common links. Under PAC, the routing information of the working paths of two connections, which includes lightpath routing information, is needed to decide whether the backup paths of these two connections can share backup capacity along common lightpaths (in the case of MPAC) or common links (in the case of SPAC).

From control point of view, PAL has lower signaling overhead. Assume that a lightpath can carry up to g connections. (In today's networks, g is typically 192 since wavelength capacity is STS-192 and the lowest bandwidth granularity is typically STS-1.) When a link fails, W lightpaths can be disrupted in the worst case. In PAL, at most W protection-switching processes are needed. However, in PAC, up to $W \times g$ protection-switching processes are required in the worst case. As protection-switching processes for shared protection typically require signaling, PAL demands lower control bandwidth and involves lower signaling complexity compared to PAC.

6.3 Heuristic Algorithms

Since it is $\mathcal{NP}$-complete to provision a connection request under the current network state with shared protection, we develop heuristics for MPAC, SPAC, and PAL in this section.

6.3.1 MPAC Heuristic

In response to a new connection request, MPAC computes two link-disjoint paths based on the current network state and appropriate backup-sharing measurement. Below, we elaborate the backup-sharing measurement, network-state representation, a modified k-distinct-loopless-path algorithm, and route computation.

6.3.1.1 Backup-sharing measurement

Every lightpath is associated with a conflict set[2] to identify the sharing potential between backup paths. The conflict set ν_l for lightpath l can be represented as an integer set, $\{\nu_l^e | \ \forall e \in E, 0 \leq \nu_l^e \leq \text{STS-192}\}$, where ν_l^e represents the amount of traffic that will be rerouted on lightpath l when link e fails. The amount of backup capacity reserved on lightpath l is thus $\nu_l^* = \max_{\forall e}\{\nu_l^e\}$. The difference $\nu_l^* - \nu_l^e$ indicates the potential "free" capacity for backing up a new working path traversing link e (and the corresponding backup path traverses lightpath l).

The union of the conflict sets for all the lightpaths aggregates the per-connection-based backup-sharing information, and the size of the conflict set depends only on the number of lightpaths and the number of links, not on the number of connections. In the absence of a mechanism such as the conflict set, per-connection-based information is necessary for identifying shareable backup capacity [Bouillet et al., 2002a]. Thus, it is advantageous to use conflict set since the number of connections can be significantly more than the number of lightpaths.

6.3.1.2 Grooming-node modeling and network-state representation

Under the current network state, a connection request may be carried by existing lightpaths, by newly-established lightpaths (based on available wavelengths and free grooming ports), or by both existing lightpaths and newly-setup lightpaths. While the graph defined in Section 6.1 takes into account wavelength constraints, the graph does not accommodate existing lightpath information. Moreover, grooming-port constraints apply if a connection is to be carried by both existing lightpaths and newly-established lightpaths. Therefore, a more powerful mechanism—which can accommodate wavelength constraints, grooming-port constraints, and existing lightpath information—is needed to represent the network state and to facilitate route computation.

We adopt the generic graph model in [Zhu et al., 2003a] to represent the network state as an auxiliary graph. For our grooming-node architecture in Fig. 5.1, W-Fabric is modeled as the λ layer consisting of input vertex[3] λ_I and output vertex λ_O; G-Fabric is modeled as the access layer consisting of input vertex A_I and output vertex A_O; grooming-add port is modeled by an edge

[2] The conflict set defined here is related to the conflict vector in [Mohan et al., 2001], the aggregated square matrix in [Liu et al., 2001], and the "bucket" link metric in [Su and Su, 2001b, Su and Su, 2001a].

[3] For clarity, we refer to node and link in the auxiliary graph as vertex and edge.

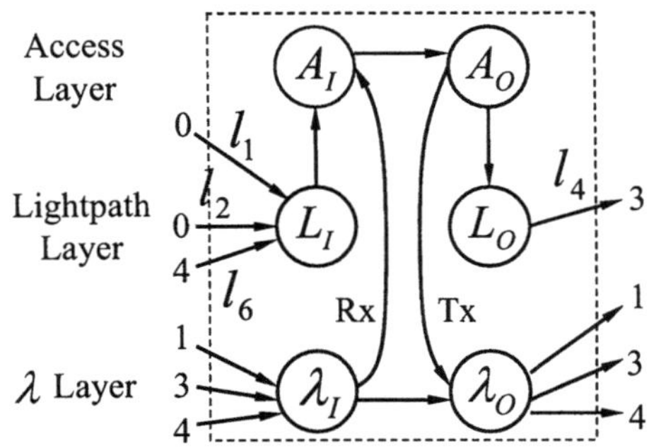

Figure 6.5. Graph representation of node 2 in Fig. 6.3(c).

from vertex A_O to vertex λ_O; and grooming-drop port is modeled by an edge from vertex λ_I to vertex A_I. A unidirectional fiber is represented as an edge from vertex λ_O at the source node to vertex λ_I at the destination node of the lightpath. A lightpath layer consisting of input vertex L_I and output vertex L_O is added to model existing lightpaths sourced/sunk at a node. A lightpath is represented as an edge from vertex L_O at the source node to vertex L_I at the destination node. Every edge is associated with two attributes: one indicating the available capacity and the other indicating the cost of the resource which the edge represents.

As an example, the state of node 2 in Fig. 6.3(c) is modeled in Fig. 6.5. For the four auxiliary edges—$\langle \lambda_I, \lambda_O \rangle$, $\langle L_I, A_I \rangle$, $\langle A_I, A_O \rangle$, and $\langle A_O, L_O \rangle$—the capacity is infinity and the cost is zero. The available capacity of any other edge e is the available capacity of the resource which edge e represents, e.g., the free capacity of edge $\langle \lambda_I, A_I \rangle$ (in gray) is zero since $R = 0$ for node 2. The grayed edge from vertex λ_O at node 4 (not shown in this figure) to vertex λ_I at node 2 indicates there are no free wavelength on link $\langle 4, 2 \rangle$. The cost of a lightpath edge is the sum of the costs of the links which the lightpath traverses. The cost of lightpath edge l_1 is two if we assume unity link cost (multiple edges between the same vertex pair is distinguished by unique sequence numbers).

By modeling every grooming node as above, the current network state—which includes wavelength usage, grooming-port usage, and available lightpath capacity—can be represented as one auxiliary graph.

6.3.1.3 Route computation

Based on the auxiliary network-state graph and the backup-sharing measurement, MPAC computes two link-disjoint paths for a connection request. The basic idea of MPAC is to judiciously enumerate paths since joint optimization of the working and backup paths may not be possible due to the NP-completeness of the problem. A formal specification of MPAC is shown in Algorithm 6.1. Further elaboration follows.

MPAC enumerates K candidate working paths in the auxiliary graph based on the customized algorithm for computing K distinct loopless paths in Section 6.3.1.4 below. For each candidate working path, MPAC computes a SRLG-disjoint minimal-cost path as backup based on cost function $C_{\text{MPAC}}(e)$. (The cost function $C_{\text{MPAC}}(e)$ is similar to the cost functions used in some previous work [Doshi et al., 1999, Mohan et al., 2001, Su and Su, 2001b, Li et al., 2002, Bouillet et al., 2002b], but it is customized in our context to accommodate grooming constraints.) In $C_{\text{MPAC}}(e)$, ϵ is an infinitesimal constant such as 10^{-9}. Then, MPAC selects the path pair of minimal cost. Please note that, in Step 3 of Algorithm 6.1, the amount of resources—especially grooming ports—used by path l_w^k should be temporarily reflected to the auxiliary graph when computing path l_b^k because the computation of path l_b^k depends on the availability of grooming ports at every node.

In Algorithm 6.1, the cost of a lightpath l is the sum of the cost of the links lightpath l traverses; the cost of a working path l_w is the sum of the cost of the lightpaths path l_w traverses times the bandwidth granularity of path l_w; and the cost of a backup path l_b is the sum of the cost of the edges l_b traverses (the cost of an edge is defined by the cost function $C_{\text{MPAC}}(e)$).

Given an eligible working and backup path pair $\langle l_w, l_b \rangle$, further optimization is possible, as we have shown in [Ou et al., 2003]. However, the improvement in performance is marginal and the increase in computational complexity is remarkable in the case when connections can have different bandwidth requirements. Thus, we do not consider such optimization further. Instead, we shall investigate how to enumerate K appropriate candidate working paths below.

6.3.1.4 Modified algorithm for computing K distinct loopless paths

The purpose of enumerating paths is to explore the search space of working and backup paths as much as possible so as to compute working and backup paths close to optimum. To this end, any two of the K candidate working paths should be "*distinct*" in the sense that two equal-cost candidate working paths should lead to different backup paths. Without grooming-port constraints, any algorithm for computing K-shortest loopless paths will be sufficient. With grooming-port constraints, some important modifications are needed, as shown below.

The K-shortest loopless paths from vertex s to vertex d are the 1st, 2nd, $\cdots$, and k^{th} least-cost paths, each of which traverses a vertex at most once. One of the algorithms is due to Yen in [Yen, 1971], and it works as follows (if less than K paths exist, the algorithm will find all the K' eligible paths, $0 \leq K' < K$). Initialize the candidate path set with a shortest path (e.g., computed by Dijkstra's algorithm), and set the K-shortest loopless path list L as empty. Repeat the

Algorithm 6.1 MPAC.

Input: $G = (V, E, C, \lambda, P)$, $c = \langle s, d, B, t_h \rangle$, existing lightpath information, wavelength usage, grooming-port usage, and K.

Output: SRLG-disjoint working and backup paths, or NULL if no such paths are found.

1 construct the auxiliary graph G_s to represent the current network state (including existing lightpath information, wavelength usage, and grooming-port usage) as shown in Section 6.3.1.2,

2 compute K minimal-cost paths $L_w = \{l_w^k \mid 1 \leq k \leq K', 0 \leq K' \leq K\}$ in G_s from the access-layer output vertex of node s to the access-layer input vertex of node d based on the modified algorithm for computing k distinct loopless paths in Section 6.3.1.4 subject to the constraint that every hop along a path should have at least B units of free capacity (if there are less than K paths between the vertices, then the algorithm will compute all the K', $0 \leq K' < K$, eligible paths); return NULL if L_w is empty,

3 for each candidate working path l_w^k in L_w, compute a minimal-cost path l_b^k from the access-layer output vertex of node s to the access-layer input vertex of node d based on the following edge-cost function $C_{\text{MPAC}}(e)$:

(a) if edge e does not represent a lightpath or a link, then $C_{\text{MPAC}}(e) = C(e)$,

(b) if edge e represents a link f, then

$$C_{\text{MPAC}}(e) = \begin{cases} +\infty & \text{if } e \text{ is not SRLG-disjoint with } l_w^k \text{, or link } f \text{ does not have any free wavelength} \\ B \cdot C(e) & \text{otherwise} \end{cases}$$

(c) if edge e represents a lightpath l, then

$$C_{\text{MPAC}}(e) = \begin{cases} +\infty & \text{if lightpath } l \text{ is not SRLG-disjoint with } l_w^k \text{, or } (\nu_l^* - \nu_l^{e'}) \text{ plus the free capacity of } l \text{ is less than } B \text{ for some link } e' \text{ used by any lightpath that } l_w^k \text{ traverses} \\ \epsilon & \text{if lightpath } l \text{ is SRLG-disjoint with } l_w^k \text{, and } (\nu_l^* - \nu_l^{e'}) \text{ is at least } B \text{ for every link } e' \text{ used by any lightpath that } l_w^k \text{ traverses} \\ B' \cdot C_l & \text{otherwise, where } C_l \text{ is the cost of } l \text{, and } B' = B - \min\{\nu_l^* - \nu_l^{e'}\} \text{ over all the links } e' \text{ used by any lightpath that } l_w^k \text{ traverses} \end{cases}$$

4 select the path pair $\langle l_w^k, l_b^k \rangle$ of minimal cost; return NULL if no such path pair exists,

5 allocate proper resources and update network state according to the paths l_w^k and l_b^k (if necessary): update the free capacity of lightpaths involved in the paths l_w^k and l_b^k; set up new lightpaths (consume new wavelengths and free grooming ports) according to the lightpath-setup strategy shown in Section 5.6.3 in the previous chapter,

6 for every lightpath l that l_b^k traverses, $\nu_l^{e'} \leftarrow \nu_l^{e'} + B$ for every link e' used by any lightpath that l_w^k traverses, and

7 return l_w^k as working path and l_b^k as backup path.

following steps until L has K paths or the candidate path set is empty. First, pick the minimal-cost path p from the candidate path set and append p to L. Then, for every vertex u ($u \neq d$) along path p, compute a shortest path (referred to as *spur*) from vertex u to vertex d subject to the constraints that: (1) the spur path cannot traverse any vertex along the *root*—the segment from vertex s to vertex u along path p—except vertex u, i.e., a loopless path; (2) the spur path cannot branch from vertex u on any link used by any path in L with the same root, i.e., a new path. If a spur is found, concatenate the root and the spur as a new path p' and put p' into the candidate path set. Yen's algorithm has a computational complexity of $O(K \cdot N^3)$, where N is the number of vertices.

If we directly apply Yen's algorithm to the auxiliary graph, some of the K candidate working paths are very likely to be non-distinct. Consider the following example. Denote the access-layer input (output) vertex of node u as A_I^u (A_O^u) and the λ-layer input (output) vertex of node u as λ_I^u (λ_O^u). Suppose that one of the candidate working paths, l_w, traverses edge $\langle \lambda_I^u, \lambda_O^u \rangle$, i.e., l_w is of the form $\langle A_O^s, \cdots, \lambda_I^u, \lambda_O^u, \cdots, A_I^d \rangle$. If we replace edge $\langle \lambda_I^u, \lambda_O^u \rangle$ by segment $\langle \lambda_I^u, A_I^u, A_O^u, \lambda_O^u \rangle$, then the new path $l_w' = \langle A_O^s, \cdots, \lambda_I^u, A_I^u, A_O^u, \lambda_O^u, \cdots, A_I^d \rangle$ has the same cost as path l_w (assuming that node u has free grooming-add and grooming-drop ports). One can easily verify that l_w and l_w' lead to the same backup path. To prevent such a situation, when we consider a vertex v along the current path p, we should not compute a spur from vertex v to vertex d if v and the next vertex along p are both induced by the same network node. We then repeat this process for the next vertex along path p.

The above situation is due to the construction of the auxiliary graph[4], which is introduced to accommodate the grooming-port constraints. The following scenario is introduced directly by the grooming-port constraints. Denote the lightpath-layer input (output) vertex of node u as L_I^u (L_O^u). Suppose that one of the candidate working paths, l_w, traverses a lightpath l_{uv} from node u to node v, and suppose that there is another lightpath l_{uv}' of sufficient free capacity from node u to node v following exactly the same physical path as l_{uv} does. Clearly, a new path l_w' formed through replacing l_{uv} by l_{uv}' has the same cost as l_w, and l_w' is not distinct (under our definition) with l_w. We modify the second constraint of computing a spur in Yen's algorithm to prevent such a scenario. Let L_u be the set of edges sourced at node u and used by some path in L with the same root. For any edge e in L_u representing some lightpath l, identify as set L_e all the other edges which represent lightpaths having the same physical path as l does. Let $L_u' = \cup L_e$ for any lightpath edge e in L_u. The original constraint in Yen's algorithm is that the spur path cannot use any edge in L_u. Now, the modified constraint is that the spur path cannot use any edge in $L_u \cup L_u'$.

Our final modification to Yen's algorithm is for achieving loopless paths in the auxiliary graph. Since a network node induces six vertices in the auxiliary graph (please see Fig. 6.5.), if a spur cannot traverse one of the induced vertices, then the spur should not traverse any of the other five induced vertices as well in the first constraint when computing a spur. Based on the above modifications,

[4]We remark that, if we apply Yen's algorithm to the original graph G instead of the auxiliary graph, then the grooming constraints may not be accommodated without appropriate mechanisms such as the graph model. Please refer to [Zhu and Mukherjee, 2002a, Zhu et al., 2002] for more details.

the resultant K distinct loopless paths may not be the K least-cost paths, but they explore the search space of the working and backup paths better.

6.3.1.5 Computational complexity

The computational complexity of Algorithm 6.1 is $O(E \cdot W + K \cdot N^3)$, where N is the number of network nodes, E is the number of links, W is the number of wavelengths, and K is the number of distinct paths. Specifically, the complexity of Step 1 is $O(E \cdot W)$ since there can be as many as $E \cdot W$ lightpaths; the complexity of Step 2 is $O(K \cdot N^3)$; the complexity of Step 3 is $O(K \cdot N^2)$ (the complexities of Steps 3a, 3b, and 3c are $O(1)$, $O(N)$, and $O(N^2)$, respectively); and the complexities of Steps 4, 5, 6, and 7 are $O(K)$, $O(N)$, $O(N^2)$, and $O(1)$, respectively.

6.3.2 SPAC Heuristic

Our heuristic for SPAC is similar to the heuristic for MPAC except for the backup-sharing measurement. Grooming node and network state are modeled the same way as shown in Section 6.3.1.2. Below, we elaborate on the backup-sharing measurement and the difference in route computation.

6.3.2.1 Backup-sharing measurement

Every link is associated with a conflict set to identify the sharing potential between backup paths. The conflict set ν_e for link e can be represented as an integer set, $\{\nu_e^{e'} \mid \forall e' \in E, 0 \leq \nu_e^{e'} \leq \text{STS-192} \times \lambda(e)\}$, where $\nu_e^{e'}$ represents the amount of traffic that will be rerouted on link e when link e' fails. The amount of backup capacity reserved on link e is thus $\nu_e^* = \max_{\forall e'}\{\nu_e^{e'}\}$. The difference $\nu_e^* - \nu_e^{e'}$ indicates the potential "free" capacity for backing up a new working path traversing link e' (and the corresponding backup path traverses link e). The number of wavelengths which need to be reserved on link e is $\lambda_e^b = \lceil \frac{\nu_e^*}{\text{STS-192}} \rceil$. The number of grooming-add ports at the upstream node (and the number of grooming-drop ports at the downstream node) of link e to be reserved is also λ_e^b.

6.3.2.2 Route computation

In response to a connection request, SPAC computes link-disjoint working and backup paths in a way similar to Algorithm 6.1 except for the cost function in Step 3 and the conflict-set update in Step 6. The edge-cost function for computing l_b^k (Step 3 in Algorithm 6.1) is $C_{\text{SPAC}}(e)$, defined as follows:

1 if edge e does not represent a lightpath or a link, $C_{\text{SPAC}}(e) = C(e)$,

2 if edge e represents a lightpath, then $C_{\text{SPAC}}(e) = +\infty$,

3 if edge e represents a link f, then: (i) $C_{\text{SPAC}}(e) = +\infty$ if link f is not SRLG-disjoint with l_w^k; or (STS-192 $\times \lambda_f^b - \nu_f^{e'}$) is less than B for some link e' used by any lightpath which l_w^k traverses and either link f does not have any free wavelength, or the upstream node of link f does not have any free grooming-add port, or the downstream node of link f does not have any free grooming-drop port. (ii) $C_{\text{SPAC}}(e) = \epsilon$ if link f is SRLG-disjoint with l_w^k, and $\nu_f^* - \nu_f^{e'} \geq B$ for every link e' used by any lightpath that l_w^k traverses. (iii) Otherwise $C_{\text{SPAC}}(e) = B' \cdot C(f)$, where $B' = B - \min\{\nu_e^* - \nu_e^{e'}\}$ over all the links e' used by any lightpath that l_w^k traverses.

Step 6 in Algorithm 6.1 is modified as follows. For any link e that path l_b^k traverses, compute $\nu_e^{e'} \leftarrow \nu_e^{e'} + B$ for every link e' used by any lightpath that l_w^k utilizes, update ν_e^*, and recompute λ_e^b. Reserve one more grooming-add port at the upstream node of link e and one more grooming-drop port at the downstream node of link e if λ_e^b increases by one (λ_e^b will increase by at most one).

6.3.3 PAL Heuristic

Upon the arrival of a connection request, PAL computes a survivable route—a sequence of p-lightpaths—based on the current network state and appropriate backup-sharing measurement, elaborated below.

6.3.3.1 Backup-sharing measurement

We associate a conflict set with a link to identify the sharing potential between backup paths. The conflict set ν_e for link e can be represented as an integer set, $\{\nu_e^{e'} \mid \forall e' \in E, 0 \leq \nu_e^{e'} \leq \lambda(e)\}$, where $\nu_e^{e'}$ specifies the number of working lightpaths that traverse link e' and are protected by link e, i.e., their corresponding backup paths traverse link e. The number of wavelengths reserved for backup paths on link e is thus $\nu_e^* = \max\limits_{\forall e'}\{\nu_e^{e'}\}$. By definition, there is no need to increase the number of reserved wavelengths on link e for protecting up to $\nu_e^* - \nu_e^{e'}$ more working lightpaths traversing link e'.

6.3.3.2 Network-state representation

The current network state is collectively represented by the set of existing p-lightpaths, the conflict set $\nu = \{\nu_e | e \in E\}$, and a digraph $G = (V, E, C, \lambda', P_T', P_R')$ (where $\lambda' : E \rightarrow Z^+$ specifies the number of available wavelengths on each link, and $P_T'(P_R') : V \rightarrow Z^+$ specifies the number of available grooming-add (grooming-drop) ports at a node). Please note that the detailed routing information about a p-lightpath is not needed for route computation since the conflict set has already aggregated that information. The only information about a p-lightpath needed for route computation is the source node, the destination node, and the available bandwidth of a p-lightpath. We will use NS to denote network state below.

6.3.3.3 Route computation

The basic idea of PAL is to extend a standard shortest-path algorithm such that every hop along the resultant shortest path corresponds to a p-lightpath, which can be either an existing p-lightpath of sufficient free capacity or a new p-lightpath consisting of fresh wavelength links and free grooming ports. The challenge here is to accommodate the backup sharing between the existing p-lightpaths and the new p-lightpaths and the backup sharing among the new p-lightpaths while computing the survivable route. PAL is related to the link-protection algorithm in [Kodialam and Lakshman, 2001] and works as follows.

In order to keep track of backup sharing and to maintain virtual adjacency[5] relationship, we associate to every node $u \in V$ a network state NS(u). NS(u) represents the updated network state after the survivable route from node s to node u is set up based on the current network state. We also associate to every node $u \in V$ a cost C_u, which represents the cost of the survivable route from node s to node u.

For any two nodes $u, v \in V$, to decide whether node u is node v's previous node along the survivable route from the source node s to the destination node d, we consider the following two possibilities. First, if an existing p-lightpath l_{uv}^e from node u to node v of sufficient free capacity exists and C_u plus the cost of l_{uv}^e's working lightpath is less than C_v, then node u is chosen as node v's previous node, p-lightpath l_{uv}^e is chosen as node u's previous hop, C_v is updated as the sum of C_u and the cost of l_{uv}^e's working lightpath, and NS(v) is the same as NS(u) except that the free capacity of p-lightpath l in NS(v) is reduced by B. Second, if an existing eligible p-lightpath from node u to node v does not exist, we check if a new one can be set up based on the network state in NS(u). If a new p-lightpath l_{uv}^n can be set up and C_u plus the cost of l_{uv}^n is less than C_v, then node u is chosen as node v's previous node, p-lightpath l_{uv}^n is chosen as node u's previous hop, C_v is updated as the sum of C_u and the cost of l_{uv}^n, and NS(v) is the network state after setting up l_{uv}^n in NS(u).

The above procedure is executed similarly to a shortest-path algorithm until C_d reaches its minimum. Once C_d reaches its minimum, the survivable route from node s to node d can be retrieved by backtracking along the previous hop, starting from node d.

The key aspect of our PAL heuristic is that the tentative changes (in network state) introduced by the survivable route from the source node s to any other node u is captured by the network state at node u. As a result, to decide

[5]Node v is referred to be virtually adjacent to node u if there exists a p-lightpath from node u to node v.

Algorithm 6.2 PAL.

Input: $G' = (V, E, C, \lambda', P_T', P_R')$, $\nu = \{\nu_e | e \in E\}$, $c = \langle s, d, B, t_h \rangle$, existing p-lightpath information, wavelength usage, grooming-port usage, and K.
Output: A concatenation of p-lightpaths from node s to node d of free capacity no less than B, or NULL.

1 **Initialization**
$V' \leftarrow V$; $C_u \leftarrow +\infty, \text{NS}(u) \leftarrow \text{NULL}, \forall u \in V' - \{s\}$; $C_s \leftarrow 0, \text{NS}(s) \leftarrow$ Current network state

2 **Iteration**
while ($d \in V'$) do

$u \leftarrow \arg \min_{u \in V'} \{C_u\}$, $V' \leftarrow V' - \{u\}$

if ($C_u = +\infty$) return NULL

for each node $v \in V'$ do

PAL_RELAX_NEXT_HOP(u, v, B) (Algorithm 6.3)

3 **Post-process:** retrieve the path and update network state
if ($C_d = +\infty$) then return NULL; otherwise

retrieve the survivable route by following the previous hop starting from node d

update the current network state according to NS(d)

return the survivable route

---

whether a new p-lightpath l_{uv}^n from node u to any other node v can be set up, the backup sharing between existing p-lightpaths and the new p-lightpath l_{uv}^n is accommodated since the conflict set ν of NS(u) measures the backup-sharing potential. Furthermore, the backup sharing among new p-lightpaths is also accommodated because the survivable route from node s to node u can have new p-lightpaths as intermediate hops. In that case, the sharing potential introduced by these new p-lightpaths (used by the survivable route from node s to node u) is reflected in the conflict set ν of NS(u). Since the computation of p-lightpath l_{uv}^n is based on NS(u), the above forms of backup sharing are correctly captured. A formal specification of PAL is shown in Algorithm 6.2.

6.3.3.4 Computational complexity

The complexity of Algorithm 6.2 is $O(E \cdot W + K \cdot N^5)$, where E is the number of links, W is the number of wavelengths, K is the number of alternate paths, and N is the number of nodes. Specifically, the complexity of Step 1 is $O(E \cdot W)$ since there can be as many as $E \cdot W$ lightpaths; the complexity of Step 2 is $O(K \cdot N^5)$ since Algorithm 6.3 is executed $O(N^2)$ times and the complexity of Algorithm 6.3 is $O(K \cdot N^3)$ (in particular, the complexities of

Algorithm 6.3 PAL_RELAX_NEXT_HOP(u, v, B).

1 let both p-lightpaths l_{uv}^{e} and l_{uv}^{n} be NULL

2 let L_{uv} be the set of existing p-lightpaths from node u to node v of free capacity no less than B. Let l_{uv}^{e} be the p-lightpath of minimal working-lightpath cost among all the p-lightpaths in L_{uv}

3 if l_{uv}^{e} is not NULL and C_u plus the cost of l_{uv}^{e}'s working lightpath is less than C_v, then

$C_v \leftarrow C_u$ plus the cost of l_{uv}^{e}'s working lightpath

NS(v) $\leftarrow$ NS(u), and reduce the available bandwidth of l_{uv}^{e} by B in NS(v)

set l_{uv}^{e} as node v's previous hop

4 if l_{uv}^{e} is NULL and node u has free grooming-add ports and node v has free grooming-drop ports then execute the following steps *based on NS(u)*

$L_{uv}^{n} \leftarrow \emptyset$

compute K minimal-cost paths $\{l_{uv}^{k} \mid 1 \leq k \leq K', 0 \leq K' \leq K\}$ from node u to node v using fresh wavelength links based on the original Yen's algorithm

for each path $l_{uv}^{k} \in \{l_{uv}^{k} \mid 1 \leq k \leq K', 0 \leq K' \leq K\}$ do

compute a minimal-cost path $\hat{l}_{uv}^{k}$ based the following link-cost function $C_{\text{PAL}}(e)$:

(i) $C_{\text{PAL}}(e) = +\infty$ if link e is not SRLG-disjoint with l_{uv}^{k}, or link e does not have any free wavelength and ν_e^* is equal to $\nu_e^{e'}$ for some link e' that l_{uv}^{k} traverses.

(ii) $C_{\text{PAL}}(e) = \epsilon$ if link e is SRLG-disjoint with l_{uv}^{k} and $\nu_e^{e'} < \nu_e^*$ for every link e' that l_{uv}^{k} traverses.

(iii) Otherwise, $C_{\text{PAL}}(e) = C(e)$.

$L_{uv}^{n} \leftarrow L_{uv}^{n} \cup \{\langle l_{uv}^{k}, \hat{l}_{uv}^{k} \rangle\}$ if $\hat{l}_{uv}^{k}$ exists

let l_{uv}^{n} be the p-lightpath of minimal cost in L_{uv}^{n}

5 if l_{uv}^{n} is not NULL and C_u plus the cost of l_{uv}^{n} is less than C_v, then

$C_v \leftarrow C_u$ plus the cost of l_{uv}^{n}

NS(v) $\leftarrow$ NS(u), and update NS(v) as follows: reduce the number of free grooming-add ports at node u by one; reduce the number of free grooming-drop ports at node v by one; consume fresh wavelength links along l_{uv}^{n}'s working lightpath; and update ν: $\nu_e^{e'} \leftarrow \nu_e^{e'} + 1$ for any link e along l_{uv}^{n}'s backup path and any link e' along l_{uv}^{n}'s working path

set l_{uv}^{n} as node v's previous hop

Steps 1 to 5 of Algorithm 6.3 are $O(1)$, $O(E \cdot W)$, $O(E \cdot W)$, $O(K \cdot N^3)$, and $O(E \cdot W)$, respectively); and the complexity of Step 3 is $O(E \cdot W)$.

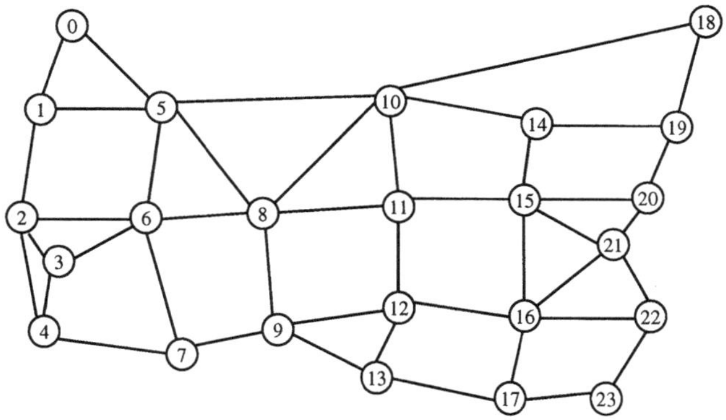

Figure 6.6. A 24-node example network topology.

6.4 Illustrative Numerical Results

We simulate a dynamic network environment with the assumptions that the connection-arrival process is Poisson and the connection-holding time follows a negative exponential distribution. For the illustrative results shown here, the capacity of each wavelength is STS-192; the number of the connection requests follows the distribution STS-1 : STS-3c : STS-12c : STS-48c : STS-192c = 300 : 20 : 6 : 4 : 1 (which is close to the bandwidth distribution in a practical backbone network); connection requests are uniformly distributed among all node pairs; average connection-holding time is normalized to unity; the cost of any link is unity; load (in Erlang) is defined as connection-arrival rate times average holding time times a connection's average bandwidth normalized in the unit of STS-192; and our example network topology with 16 wavelengths per fiber is shown in Fig. 6.6. 100,000 connections were simulated in each experiment. The value of ϵ in the cost function is set to 10^{-6}; ϵ can trade off backup sharing and backup-path length, as shown in [Bouillet et al., 2002b].

The number of grooming ports at a node is set as the number of wavelengths times its nodal degree times a scalar Δ ($0 \leq \Delta \leq 1$, $\Delta = 1$ implies that any incoming wavelength to the W-Fabric can be dropped to the G-Fabric). The number of alternate paths K for the three schemes is two. Later in Section 6.4.4, we shall examine the effect of different values of these parameters.

We now quantitatively compare PAL to MPAC to SPAC using the following metrics: bandwidth-blocking ratio (BBR) and resource-efficiency ratio (RER).

6.4.1 Bandwidth-Blocking Ratio

BBR is defined as the amount of bandwidth blocked over the amount of bandwidth offered. Please note that pure blocking probability, defined as the percentage of the *number* of connections blocked, cannot reflect the effective-

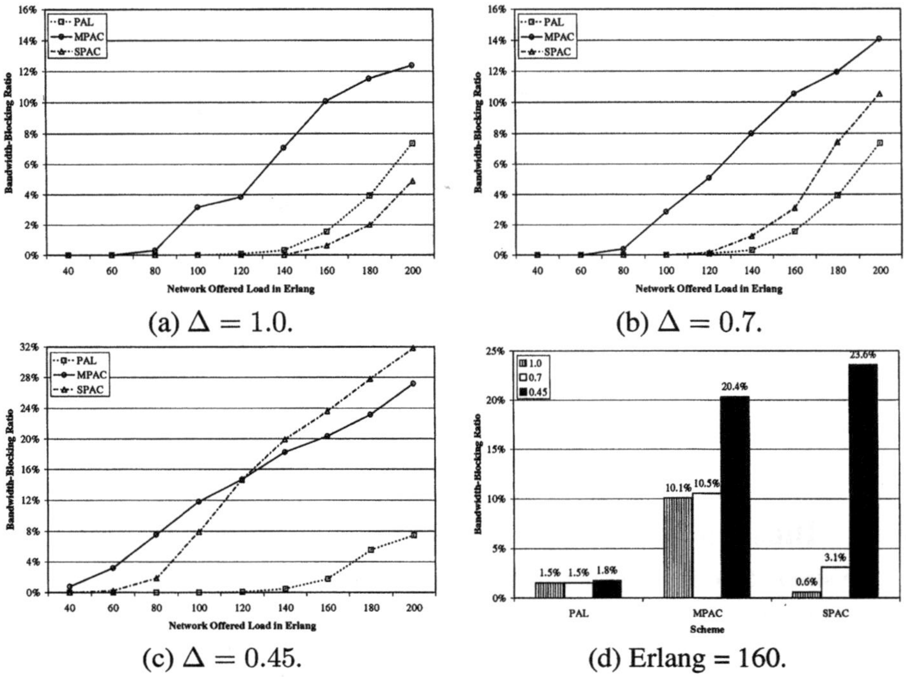

(a) $\Delta = 1.0$. (b) $\Delta = 0.7$. (c) $\Delta = 0.45$. (d) Erlang = 160.

Figure 6.7. Bandwidth-blocking ratio versus network offered load.

ness of the algorithm as connections have different bandwidth requirements. Figure 6.7 plots the BBR of the three schemes with $\Delta = 1.0, 0.7$, and 0.45. We make the following observations.

We find that PAL always has lower BBR than MPAC, and SPAC has lower BBR than MPAC when the number of grooming ports is large (e.g., $\Delta = 1.0$ and 0.7) or the number of grooming ports is small and the network offered load is moderate (e.g., $\Delta = 0.45$ and the network offered load is less than 120 Erlangs). This leads to our first observation: *It is beneficial to groom working paths and backup paths separately, as is the case in PAL and SPAC.*

Our second observation is that *SPAC has the lowest BBR when the number of grooming ports is sufficient (e.g., $\Delta = 1.0$)*, as shown in Fig. 6.7(a). This is because SPAC has the maximum freedom in backup sharing when the number of grooming ports is sufficient (please see Section 6.2.4.2).

Our third observation is that *PAL achieves the lowest BBR when the number of grooming ports is moderate or small (e.g., $\Delta = 0.7$ and 0.45)*, as shown in Figs. 6.7(b) and (c). The main reason for this is that backup paths do not consume grooming ports under PAL (Section 6.2.4.1).

Figure 6.7(d) shows the BBR of the three schemes with different values of Δ under the same network offered load, 160 Erlangs. Clearly, the decrease in

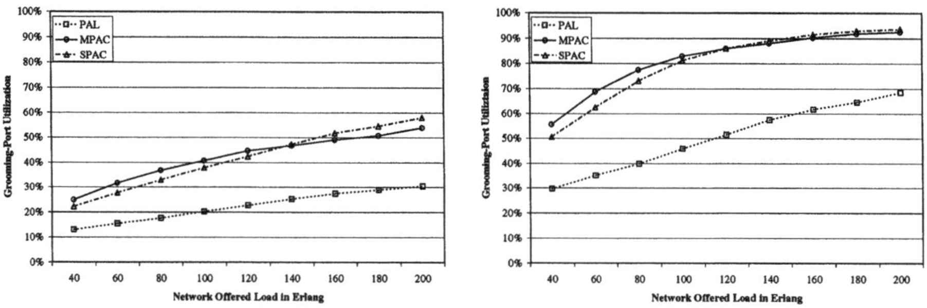

(a) $\Delta = 1.0$. (b) $\Delta = 0.45$.

Figure 6.8. Grooming-port utilization versus network offered load.

grooming capacity impacts PAL the least and SPAC the most. Again, this is because PAL trades bandwidth efficiency in routing for grooming-port savings, and SPAC trades grooming ports for the flexibility in backup sharing (Section 6.2.4). More reasons for the above observations will be further elaborated below.

6.4.2 Resource Utilization

We consider two types of resources: grooming ports and wavelengths. Grooming-port utilization is the percentage of grooming ports used. Wavelength utilization is the percentage of wavelength links utilized by lightpaths. Intuitively, lower wavelength utilization and lower grooming-port utilization (under the same BBR) are desirable. Resource utilization can differentiate blockings due to different resources. For example, when grooming-port utilization is high and wavelength utilization is moderate or low, then connections are more likely blocked due to insufficient grooming ports.

6.4.2.1 Grooming-port utilization

Figure 6.8 plots the grooming-port utilization of the three schemes for $\Delta = 1.0$ and $\Delta = 0.45$. (The plot for $\Delta = 0.7$ is similar to the one for $\Delta = 1.0$, so it is not shown here.)

Our first observation is that PAL has the lowest grooming-port utilization under different values of Δ. The reason is that backup paths consume grooming ports under both SPAC and MPAC, but not under PAL. This confirms our result in Fig. 6.7(d) and the analysis in Section 6.2.4 that PAL trades bandwidth efficiency for the savings in grooming ports. As a result, under PAL, connections are more likely blocked due to insufficient wavelengths.

Our second observation is that SPAC has lower grooming-port utilization than MPAC when the network offered load is moderate or low and SPAC has higher grooming-port utilization than MPAC when the network offered load is

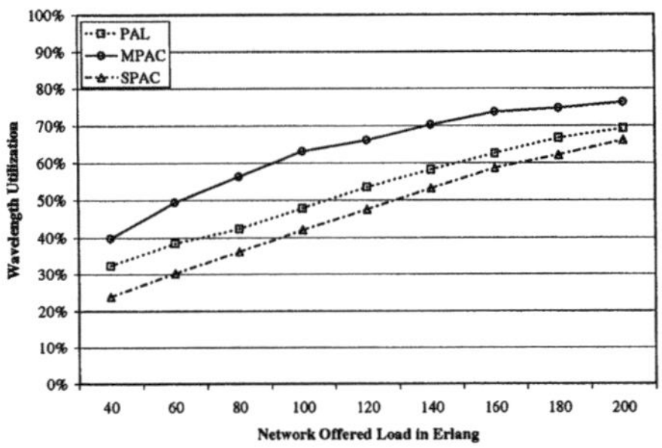

(a) $\Delta = 1.0$.

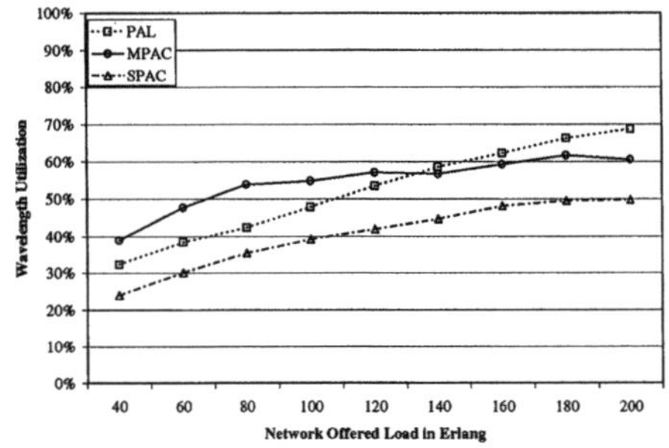

(b) $\Delta = 0.45$.

Figure 6.9. Wavelength utilization versus network offered load.

high. When the network offered load is moderate or low, the lightpaths under MPAC are only moderately loaded. That is, a substantial number of grooming ports is used by these lightpaths to carry relatively moderate network load. Therefore, MPAC has higher grooming-port utilization when the network offered load is moderate or low. When the network offered load is high, the lightpaths under MPAC are heavily loaded. Under SPAC, a significant number of grooming ports is used by backup paths since every single reserved wavelength needs a grooming-add port and a grooming-drop port. Thus, SPAC has higher grooming-port utilization when the network offered load is high.

6.4.2.2 Wavelength utilization

Figure 6.9 plots the wavelength utilization of the three schemes for $\Delta = 1.0$ and $\Delta = 0.45$. (The plot for $\Delta = 0.7$ is similar to the one for $\Delta = 1.0$.)

For $\Delta = 1.0$, MPAC has the highest and SPAC has the lowest wavelength utilization, as shown in Fig. 6.9(a). This is because SPAC and PAL have more flexibility in backup sharing (Section 6.2.4.2). Compared to PAL, SPAC has further flexibility in grooming since PAL work at an aggregated level in routing (Section 6.2.4.1). Therefore, SPAC has the lowest wavelength utilization.

For $\Delta = 0.45$, the trend of the three curves is similar to the one for $\Delta = 1.0$ when the network offered load is no more than 120 Erlangs. When the network offered load is higher than 120 Erlangs, PAL has higher wavelength utilization than MPAC. The reason is that, when the network offered load is high (e.g., more than 120 Erlangs), MPAC and SPAC have much higher BBR than PAL for $\Delta = 0.45$, as shown in Fig. 6.7(c). Many connection requests are now blocked because of a shortage of grooming ports. As a result, some wavelength links are not well utilized under MPAC and SPAC, and their wavelength utilization are lower than that of PAL.

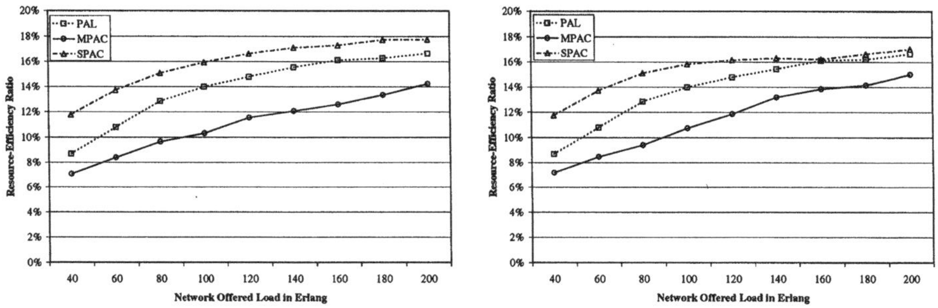

(a) $\Delta = 1.0$. (b) $\Delta = 0.45$.

Figure 6.10. Resource-efficiency ratio, $\mathcal{E}(1,0)$, versus network offered load.

6.4.3 Resource-Efficiency Ratio

6.4.3.1 Definition

To better evaluate the performance of our route-computation heuristics, we introduce a new metric, called *resource-efficiency ratio* (RER) $\mathcal{E}$, which is defined as the carried load (weighted by time and normalized to STS-192 capacity) divided by the amount of allocated resources in terms of wavelength channels and grooming ports (weighted by time). This metric is defined as follows:

$$\mathcal{E}(W_\lambda, W_g) = \frac{\sum_i \rho_i \times t_i}{W_\lambda \times \sum_i \beta_i \times t_i + W_g \times \sum_i \gamma_i \times t_i}$$

where t_i is the time period between the ith event (connection arrival or departure) and $(i+1)^{th}$ event; ρ_i is the network carried load during the time period t_i; β_i is the number of wavelength links used during t_i; γ_i is the number of grooming ports used during t_i; W_λ and W_g are the relative weight of a wavelength link versus a grooming port. (Please note that ρ_i, β_i, and γ_i do not change during time period t_i as there is no other event during the period.) Basically, $\mathcal{E}$ measures how efficiently resources have been used.

6.4.3.2 Wavelength efficiency

If $W_\lambda = 1$ and $W_g = 0$, RER $\mathcal{E}(1,0)$ measures how efficiently wavelength channels have been utilized. Figure 6.10 plots the RER $\mathcal{E}(1,0)$ for $\Delta = 1.0$ and 0.45 (the plot for $\Delta = 0.7$ is similar to the one for $\Delta = 1.0$). MPAC has the lowest wavelength efficiency since PAL and SPAC have more flexibility in backup sharing. Furthermore, SPAC has the highest wavelength efficiency because PAL works at lightpath level and lightpaths are not perfectly filled.

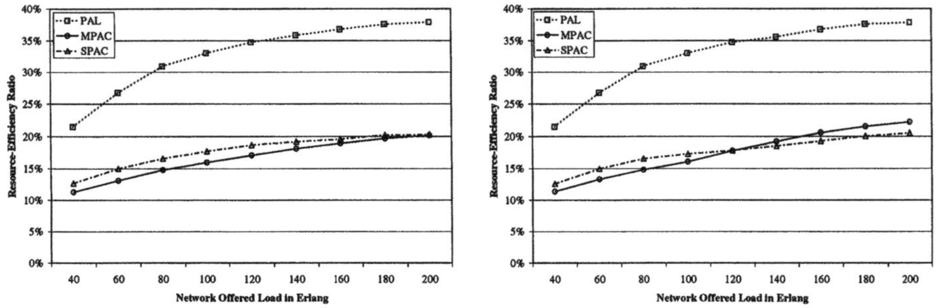

(a) $\Delta = 1.0$. (b) $\Delta = 0.45$.

Figure 6.11. Resource-efficiency ratio, $\mathcal{E}(0,1)$, versus network offered load.

6.4.3.3 Grooming-port efficiency

If $W_\lambda = 0$ and $W_g = 1$, RER $\mathcal{E}(0,1)$ measures how efficiently grooming ports have been utilized. Figure 6.11 plots the RER $\mathcal{E}(0,1)$ for $\Delta = 1.0$ and 0.45 (the plot for $\Delta = 0.7$ is similar to the one for $\Delta = 1.0$).

Our first observation is that PAL has the highest grooming-port efficiency under different values of Δ. The reason is that backup paths consume grooming ports under both SPAC and MPAC, but not under PAL. This confirms our result in Fig. 6.7(d) and the analysis in Section 6.2.4 that PAL trades bandwidth efficiency for the savings in grooming ports. As a result, under PAL, connections are more likely blocked due to insufficient wavelengths.

Our second observation is that SPAC has higher grooming-port efficiency than MPAC when the network offered load is moderate or low and SPAC has lower grooming-port efficiency than MPAC when the network offered load is high. When the network offered load is moderate or low, the lightpaths under MPAC are only moderately loaded. That is, a substantial number of grooming ports is used by these lightpaths to carry relatively moderate network load. Therefore, MPAC has lower grooming-port efficiency when the network offered load is moderate or low. When the network offered load is high, the lightpaths under MPAC are heavily loaded. Under SPAC, a significant number of grooming ports is used by backup paths since every single reserved wavelength needs a grooming-add port and a grooming-drop port. Thus, SPAC has lower grooming-port efficiency when the network offered load is high.

6.4.3.4 Tradeoff between wavelengths and grooming ports

The three schemes trade off the utilization between wavelengths and grooming ports. Below, we show that either PAL or SPAC can have the highest RER, depending on the relative weight of a wavelength channel (W_λ) and a grooming port (W_g). However, the intuitive scheme, MPAC, will not have the highest RER with any possible W_λ and W_g combination because both RER $\mathcal{E}(1,0)$ and

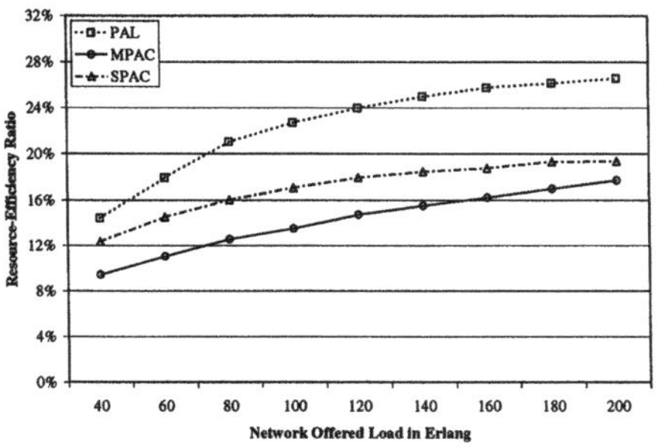

(a) $\Delta = 1.0$.

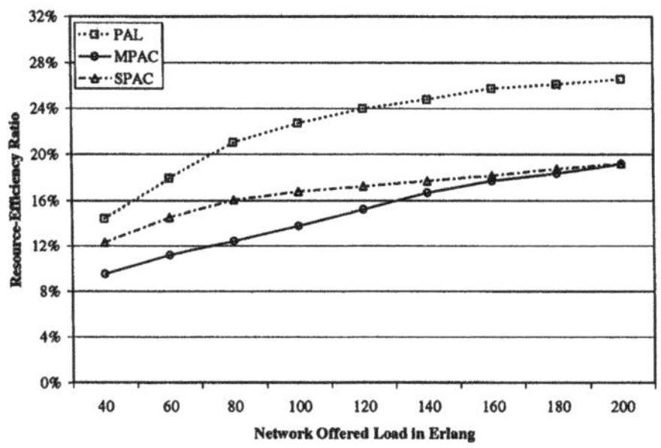

(b) $\Delta = 0.45$.

Figure 6.12. Resource-efficiency ratio, $\mathcal{E}(\frac{1}{3}, \frac{2}{3})$, versus network offered load.

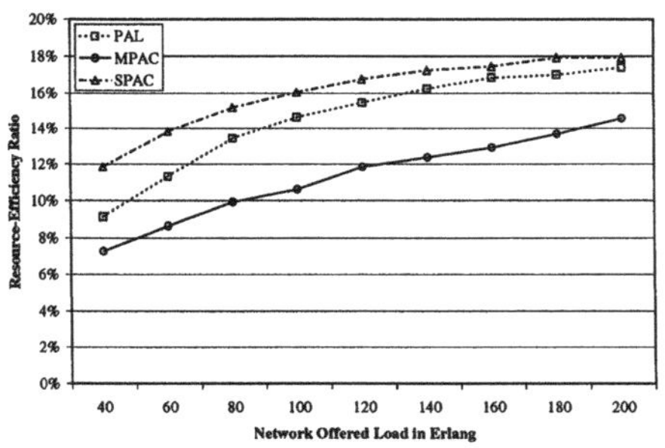

(a) $\Delta = 1.0$.

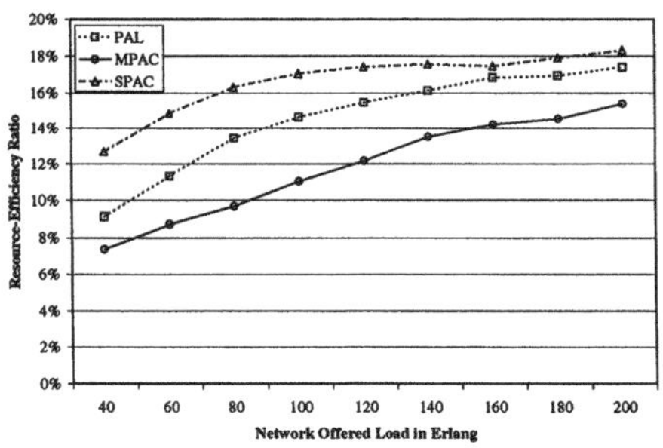

(b) $\Delta = 0.45$.

Figure 6.13. Resource-efficiency ratio, $\mathcal{E}(\frac{12}{13}, \frac{1}{13})$, versus network offered load.

RER $\mathcal{E}(0, 1)$ for MPAC are lower than those for PAL, as shown in Figs. 6.10 and 6.11. The fundamental reason for this is that MPAC has disadvantages in backup sharing compared to either PAL or SPAC (please refer to Section 6.2.4.2).

Figure 6.12 plots the RER $\mathcal{E}(\frac{1}{3}, \frac{2}{3})$ for $\Delta = 1.0$ and 0.45. We observe that PAL has the highest RER $\mathcal{E}(\frac{1}{3}, \frac{2}{3})$. In general, based on the results in Figs. 6.10, 6.11, and 6.12, PAL has the highest RER $\mathcal{E}(W_\lambda, W_g)$ when a grooming port weights more than a wavelength link, e.g., $W_\lambda : W_g = 1 : 2$.

Figure 6.13 plots the RER $\mathcal{E}(\frac{12}{13}, \frac{1}{13})$ for $\Delta = 1.0$ and 0.45. We observe that SPAC has the highest RER $\mathcal{E}(\frac{12}{13}, \frac{1}{13})$. In general, based on the results in Figs. 6.10, 6.11, and 6.13, SPAC has the highest RER $\mathcal{E}(W_\lambda, W_g)$ when a wavelength link weights significantly more than a grooming port, e.g., $W_\lambda : W_g = 12 : 1$.

6.4.4 Effects of Different Parameters

Figure 6.14 plots the impact of K, the number of distinct alternate paths, on BBR for the three schemes. For MPAC and SPAC, when K increase from

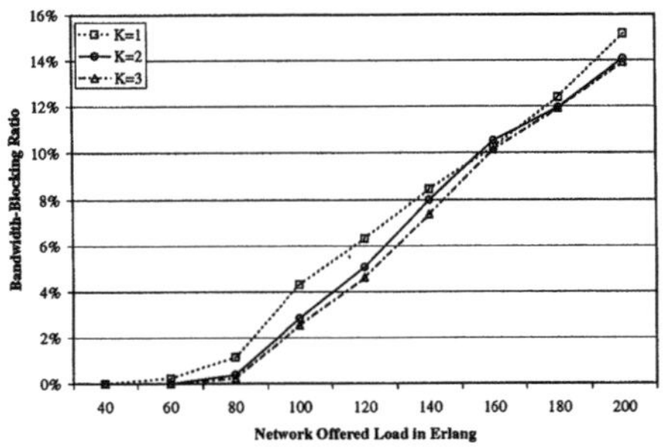

(a) MPAC, $\Delta = 0.7$.

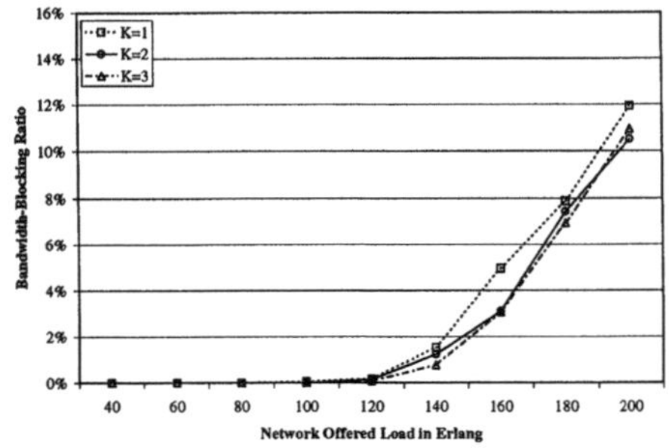

(b) SPAC, $\Delta = 0.7$.

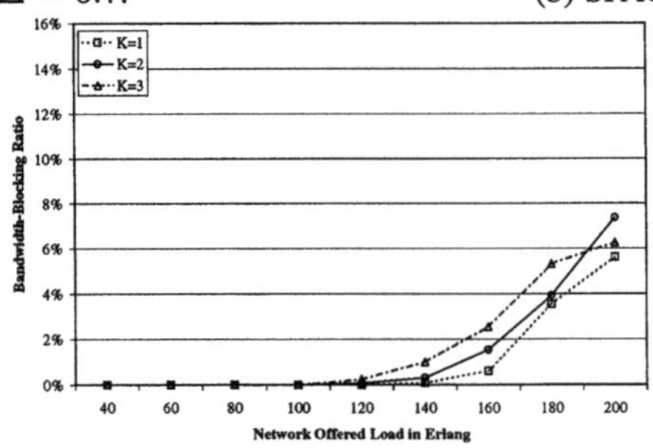

(c) PAL, $\Delta = 0.7$.

Figure 6.14. BBR versus network offered load with $K = 1, 2$, and 3.

one to two, we observe a modest reduction in BBR; when K further increases from two to three, the decrease in BBR is marginal or none. This is expected since larger K implies larger search space for MPAC and SPAC. However, BBR for PAL increases when K increases. This is because we can only apply the K-shortest-path algorithm to compute new p-lightpaths, but not to compute the final survivable route. Since the cost of an existing p-lightpath is defined as the cost of its working path to encourage the use of existing p-lightpaths, increasing K does more harm than good because larger K basically prefers the use of new p-lightpaths in PAL.

6.5 Conclusion

We investigated the survivable traffic-grooming problem for optical WDM mesh networks in a dynamic context. Based on a generic grooming-node architecture, we explored three approaches—protection-at-lightpath (*PAL*) level, mixed protection-at-connection (*MPAC*) level, and separate protection-at-connection (*SPAC*) level—for grooming a connection request with *shared* protection against single-fiber failures. Our findings are as follows. Under today's typical connection-bandwidth distribution, 1) it is beneficial to groom working paths and backup paths separately, as in PAL and SPAC; 2) separately protecting each individual connection—i.e., SPAC—yields the best performance when the number of grooming ports is sufficient; 3) protecting each specific lightpath—i.e., PAL—achieves the best performance when the number of grooming ports is moderate or small.

So far we have considered single-path routing in which a connection is routed onto one path. In the next chapter, we consider multi-path routing in which a connection can be routed onto multiple paths.

Chapter 7

SURVIVABLE VIRTUAL CONCATENATION FOR DATA OVER SONET/SDH

7.1 Introduction

7.1.1 Next-Generation SONET/SDH Technologies

SONET/SDH has historically been the dominant transport infrastructure optimized for reliable delivery of voice and private-line services in metro and backbone networks. They likely will remain in the foreseeable future as the dominant framing layer for supporting integrated data and voice services over optical transport networks to leverage the existing SONET/SDH infrastructure because of the emerging data-over-SONET/SDH (DoS) technologies: namely generic framing procedure [Hernandez-Valencia et al., 2002], virtual concatenation [ANSI T1X1.5, 2001-062, 2001, ITU, G.707, 2002], and link-capacity-adjustment scheme [ITU-T Rec. G.7042/Y.1305, 2001].

Generic framing procedure (GFP) is a traffic-adaptation protocol which maps either a physical-layer signal or a logical-link-layer signal to an octet-synchronous signal such as a SONET/SDH channel. While data traffic can be first mapped into SONET/SDH signals via GFP and then transported over a traditional SONET/SDH network, such an approach can lead to significant bandwidth inefficiency as traditional SONET/SDH provides a rigid, tiered bandwidth allocation, e.g., STS-1, STS-3c, STS-12c, STS-48c, and STS-192c. For example, a STS-48c SONET container (of approximate capacity 2.5 Gbps) is needed to carry a Gigabit data connection, such as a Gigabit Ethernet line, in such a network. This results in about 60% bandwidth wastage. Clearly, flexible bandwidth-allocation mechanisms are needed.

Virtual concatenation (VC) is one such mechanism. VC is an inverse-multiplexing technique which groups an arbitrary number of SONET/SDH containers, which may not be contiguous, to create a bigger container. A network operator can combine any number of either low-order containers (VT1.5s/VT2s in SONET) or high-order containers (STS-1s/STS-3cs in SONET), depending on the switching granularity, to create one VC group (VCG). With VC, it is now possible to provide fine granular bandwidth, e.g., multiples of VT1.5 (payload capacity about 1.536 Mbps) for low-order VC and multiples of STS-1 (payload capacity about 48.384 Mbps) for high-order VC. For example, 21 virtually concatenated STS-1s, denoted by STS-1-21v, can be used to provision a Gigabit data connection. Compared to the STS-48c bandwidth needed to carry this connection without VC, only STS-21 bandwidth is needed with VC, resulting in a huge capacity savings (of approximately 60%).

A connection carried by a VCG, referred to as a VC connection, can be inversely multiplexed onto multiple paths at the source node and merged at the destination node. Each path of a VCG, referred to as a VCG member, is routed across the network independently. As a result, VC works across traditional SONET/SDH networks because only the source and destination nodes are aware of the VCG. Since a connection can be inversely multiplexed onto multiple paths, network load can be distributed more evenly and network performance (or information-carrying capacity) can be improved. One assumption here is that the destination node can compensate the differential delay of the VCG members via buffering. Commercial SONET/SDH framer/mapper can compensate up to $\pm$125 ms differential delay of up to 64 VCG members via external random-access memory (RAM) [Stanley, 2003].

Since data traffic is bursty, it is desirable that the SONET/SDH transport network can dynamically adjust the bandwidth allocated to a connection to accommodate traffic fluctuation.

Link-capacity-adjustment scheme (LCAS) serves this purpose. LCAS is a two-way signaling protocol built on VC, and it can dynamically adjust the bandwidth of a VC connection by adding/deleting VCG members in a hitless manner (without disrupting the traffic carried on the existing VCG members). Besides, LCAS provides some degree of resilience since it can dynamically remove failed (or add recovered) VCG members. In case of network-element failures or network congestion, LCAS can reduce the bandwidth of VC connections by removing the disrupted VCG members, *thus providing degraded services, instead of "no service" at all, as in current solutions.*

As an example, consider the provisioning of a Gigabit Ethernet (GbE) connection in an optical transport network using next-generation SONET/SDH. At the source node, the GbE connection can be mapped via GFP into STS-3c-7v, each of which is routed across the SONET/SDH network independently; at the destination node, the seven independent STS-3c's are combined into a STS-3c-7v signal and mapped back to GbE frames, as shown in Fig. 7.1. If a VCG member – which is a STS-3c signal – fails, the destination node can detect the failure and notify the source node to remove the failed VCG member via LCAS. When the failure gets fixed, the destination informs the source node to add the

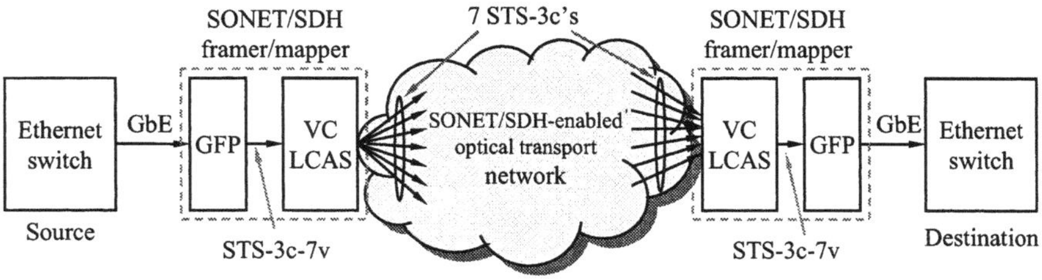

Figure 7.1. Provisioning a GbE connection in a SONET/SDH-enabled optical transport network.

recovered VCG member back into the VCG. Before the failure is fixed, the source node can also add into the VCG another path of STS-3c free capacity if it so desires and if such a path exists.

The challenge of computing routes to take advantage of VC has been an open research problem. In fact, route computation is assumed to be the responsibility of the network-management system (NMS) according to the various standards bodies. In [Zhu et al., 2003c], we quantified the benefits of using VC for unprotected traffic. In this chapter, we investigate the survivability of DoS.

7.1.2 Motivation for Survivable DoS

Data networks have limited protection capabilities. For example, depending on the size of the network, current Ethernet protection schemes based on spanning-tree algorithms can take up to dozens of seconds to converge [Metro Ethernet Forum whitepaper, 2002]. Clearly, the fault-recovery time of Ethernet is too long for voice and mission-critical applications. IP restoration is also too slow for mission-critical applications [Fumagalli and Valcarenghi, 2000, Gerstel and Ramaswami, 2000b, Sahasrabuddhe et al., 2002].

While SONET/SDH automatic protection switching (APS) provides protection against single-fiber failure within 50 ms, SONET/SDH APS is known to be resource inefficient. SONET APS incurs at least 100% (and up to 300%) backup resource overbuild [MacDonald et al., 2000]. Since a large portion of data traffic is inherently best effort and non-mission critical, providing 50-ms protection with a huge sacrifice in backup resources is not desirable.

In addition, protecting a DoS connection also differs from WDM protection without VC [Ramamurthy et al., 2003] and multiprotocol label switching (MPLS) tunnel restoration [Kodialam and Lakshman, 2003] in that one DoS connection can be inversely multiplexed onto multiple paths. With VC, multiple members of a VCG (of *one* connection) may share backup resources if these members are SRLG-disjoint; we refer to this as *intra-connection sharing*. Meanwhile, multiple members of different VCGs (of *different* connections) may share backup resources if these members are SRLG-disjoint; we refer to this as *inter-connection sharing*. Protection becomes much more complex because a flow, consisting of multiple paths, needs to be protected.

In summary, new approaches are needed for provisioning survivable DoS connections to strike an attractive balance between resource efficiency and fault-recovery time while taking advantage of the inverse-multiplexing capability of VC.

7.1.3 Our Contribution

We propose and investigate two new approaches – Protecting Individual VCG Member (PIVM) and Provisioning fast REstorable VCG (PREV) – and the associated route-computation algorithms for dynamically provisioning survivable DoS connections. Both approaches provide 100% guarantee against single-fiber failures and degraded services against multiple-fiber failures[1]. By exploiting the inverse-multiplexing capability, our approaches maximize intra-connection sharing and exploit the tradeoff between inter-connection sharing and fault-recovery time. PIVM, which is suitable for centralized implementation, achieves high backup sharing which leads to efficient resource utilization; while PREV, which is suitable for distributed implementation, achieves fast fault recovery comparable to SONET 50-ms protection (for a typical US backbone network) while still achieving modest backup sharing.

Since it may not be desirable to inversely multiplex one connection onto an arbitrary number of paths due to network control and management considerations, we also investigate the impact of VCG size, denoted by M, on network performance. We first prove that it is NP-complete to compute a minimum-cost VCG having two members, i.e., $M = 2$, of combined bandwidth no less than a given value. Then, we design an effective heuristic. Our results indicate that the network blocking performance for small values of M (e.g., $M = 2$ for some topologies) is almost as good as the case in which M is infinity.

7.1.4 Organization

The rest of this chapter is organized as follows. Sections 7.2 and 7.3 present PIVM and PREV, respectively. Section 7.4 describes our approach for controlling VCG size. Section 7.5 compares PIVM and PREV. Section 7.6 concludes this chapter.

[1]Single-fiber failures are the predominant form of failures in communication networks. Node failures are relatively rare compared to fiber failures because the switch fabric and switch-control unit in a carrier-class node are typically dedicated (1+1) protected. Therefore, nodes are assumed to be robust in this chapter.

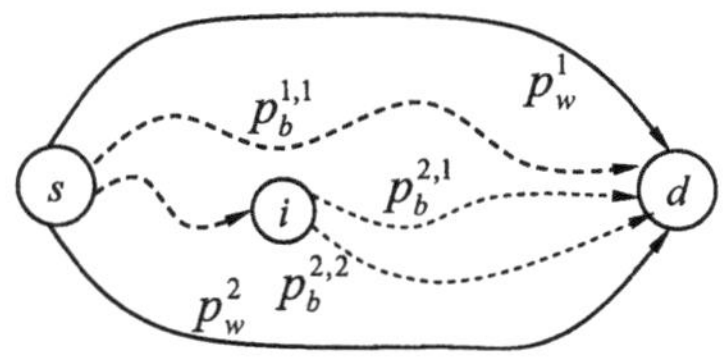

Figure 7.2. Survivable DoS approach: PIVM.

7.2 Protecting Individual VCG Member (PIVM)

7.2.1 Basic Idea

PIVM is similar to shared-mesh protection, and it exploits the inverse-multiplexing capability of VC to further increase backup sharing. Denote as $\langle s, d, u \rangle$ a connection request from node s to node d requiring u units of bandwidth; and represent the capacity of path p as $B(p)$. The basic ideas of PIVM are:

1. to route connection $\langle s, d, u \rangle$ with a working VCG $\mathcal{P}_w$ of capacity u;
2. to protect each working VCG member $p_w^k \in \mathcal{P}_w$ $(1 \leq k \leq |\mathcal{P}_w|)$ with a backup VCG $\mathcal{P}_b^k$ of capacity $B(p_w^k)$ such that any member of $\mathcal{P}_b^k$ is link-disjoint to p_w^k; and
3. to share backup resources between any two working VCG members as long as they are link-disjoint.

Please note that p_w^k and any path in $\mathcal{P}_b^l$ $(k \neq l)$ do not need to be link-disjoint. We also remark that: (1) WDM shared-mesh protection is a special case of PIVM in which $u = 1$ for every connection request; (2) MPLS tunnel protection is a special case of PIVM in which $|\mathcal{P}_w| = 1$ and $|\mathcal{P}_b^1| = 1$ $(k = 1$ because $1 \leq k \leq |\mathcal{P}_w| = 1)$ for every connection request.

7.2.2 An Example

To elaborate on the above ideas, consider the example in Fig. 7.2. The connection from node s to node d has working VCG $\mathcal{P}_w = \{p_w^1, p_w^2\}$. Working VCG member p_w^1 is protected by backup VCG $\mathcal{P}_b^1 = \{p_b^{1,1}\}$, and p_w^2 is protected by backup VCG $\mathcal{P}_b^2 = \{p_b^{2,1}, p_b^{2,2}\}$. Note that, even though VCG members $p_b^{2,1}$ and $p_b^{2,2}$ follow the same path from node s to node i, they are inversely multiplexed at node s, *not* at node i, because $p_b^{2,1}$ and $p_b^{2,2}$ are routed independently. They follow the same segment from s to i because there is enough capacity on it. Basically, $p_b^{2,1}$ and $p_b^{2,2}$ are capacity-disjoint from s to d. Backup VCG member $p_b^{1,1}$ may share backup capacity with both $p_b^{2,1}$ and $p_b^{2,2}$ if p_w^1 is link-disjoint to

p_w^2 (intra-connection sharing). $p_b^{1,1}$ may share backup capacity with any member of an existing backup VCG (not shown in the figure) if p_w^1 is link-disjoint to the corresponding working VCG member (inter-connection sharing).

In case of a working VCG member failure, e.g., p_w^2, destination d detects the failure and notifies source s. Upon receiving the notification, source s (1) removes the failed member, p_w^2, from the working VCG via LCAS; (2) signals the nodes along the backup VCG, $\mathcal{P}_b^2$, to properly configure their switches; and (3) adds all the members of the backup VCG, $\mathcal{P}_b^2$, to the working VCG via LCAS.

7.2.3 Route Computation: General Case

Before we present our route-computation approach for an incoming connection request, let us define the notations and formally state the dynamic connection-provisioning problem under PIVM constraints.

7.2.3.1 Notations

A network is represented as a weighted, directed graph $G = (V, E, C, \lambda)$, where V is the set of nodes, E is the set of unidirectional fibers (referred to as links), $C : E \rightarrow Z^+$ is the cost function for each link (where Z^+ denotes the set of positive integers), and $\lambda : E \rightarrow Z^+$ specifies the number of wavelengths on each link.

Every link is associated with a conflict set to identify the sharing potential between backup VCGs[2]. Let W_c be the wavelength capacity. The conflict set ν_e for link e can be defined as an integer set, $\{\nu_e^{e'} \mid \forall e' \in E, 0 \leq \nu_e^{e'} \leq \lambda(e') \times W_c\}$, where $\nu_e^{e'}$ represents the amount of traffic that will be rerouted on link e when link e' fails. The amount of backup capacity reserved on link e is thus $\nu_e^* = \underset{\forall e' \in E}{max}\{\nu_e^{e'}\}$. The difference $\nu_e^* - \nu_e^{e'}$ indicates the potential "free" capacity for backing up a new working VCG member which traverses link e' and whose corresponding backup VCG traverses link e.

The union of the conflict sets for all the links aggregates the per-connection-based information, and the size of the conflict set depends only on the number of links, not on the number of connections. In the absence of such a mechanism as conflict set, per-connection-based information is necessary for identifying

[2]A wavelength-routed WDM network can be wavelength continuous, in which a connection is required to occupy the same wavelength throughout its path in the network, or wavelength convertible, in which a connection may use different wavelengths on different links due to the existence of wavelength-convertion devices [Mukherjee, 1997]. In this chapter, we assume that every node is wavelength convertible because wavelength conversion comes for free due to the optical-electrical-optical (OEO) conversion at every node in current communication networks. In the wavelength-continuous case, we would associate a conflict set to a wavelength. The conflict set defined here is related to the conflict vector in [Mohan et al., 2001], the aggregated square matrix in [Liu et al., 2001], and the "bucket" link metric in [Su and Su, 2001b].

shareable backup bandwidth [Bouillet et al., 2002a]. It is, thus, advantageous to use conflict set since the number of connections can be significantly more than the number of links.

7.2.3.2 Problem Statement

We now formally state the dynamic connection-provisioning problem under PIVM constraints as follows: Given a WDM network as $G = (V, E, C, \lambda)$ and the set of existing connections (or the associated conflict set $\{\nu_e \mid e \in E\}$), for each incoming connection request $\langle s, d, u \rangle$, compute a working VCG $\mathcal{P}_w$ from node s to node d of bandwidth u and a set of backup VCGs (one for each working VCG member) under backup-sharing constraints while minimizing the total cost of the working and backup VCGs.

The existence version of the above problem for provisioning one connection request under the current network state is NP-complete. The is because the existence version of shared-mesh protection, which is a special case of this problem in which $u = 1$ for every connection request, has been proven to be NP-complete [Ou et al., 2003]. Therefore, practical heuristics are needed, as shown below.

7.2.3.3 The PIVM Heuristic

Upon the arrival of a new connection request $\langle s, d, u \rangle$, our PIVM heuristic operates as follows:

1 Compute a minimum-cost flow f_w from node s to node d of bandwidth u using a standard minimum-cost flow algorithm [Ahuja et al., 1993].

2 Extract as the working VCG $\mathcal{P}_w$ a set of paths constituent to flow f_w.

3 For every working VCG member $p_w^k \in \mathcal{P}_w$, compute as the backup VCG $\mathcal{P}_b^k$ an integral minimum-cost flow from node s to node d of bandwidth $B(p_w^k)$ while accommodating backup sharing.

The challenge here is how to take into consideration backup sharing. In WDM shared-mesh protection and MPLS tunnel restoration, we can accommodate backup sharing by simply manipulating link cost when computing a backup path. For example, if link e has a shareable wavelength, then the cost of link e can be redefined much smaller than the original cost of link e when computing the backup path [Bouillet et al., 2002b, Doshi et al., 1999, Su and Su, 2001a, Xiong et al., 2003]. Manipulating link cost does not apply here because the cost of a link depends on the amount of flow traversing the link, and we do not know in advance the amount of flow on every link.

We accommodate backup sharing by introducing parallel links. For a given working VCG member p_w^k, calculate for every link e the amount of shareable

Algorithm 7.1 PIVM.

Input: $\langle s, d, u\rangle, G = (V, E, C, \lambda), B_a : E \to Z^+$ specifying the amount of available bandwidth on every link, and the conflict set $\{\nu_e \mid e \in E\}$.
Output: A minimum-cost working VCG $\mathcal{P}_w$ and a set of backup VCGs (one for each working VCG member); otherwise NULL if no eligible solution is found.

1 Compute in G an integral minimum-cost flow f_w of bandwidth u from node s to node d based on the available bandwidth of each link; return NULL if f_w is not found.

2 Extract as the working VCG $\mathcal{P}_w$ a set of paths constituent to flow f_w and update B_a.

3 For each working VCG member $p_w^k \in \mathcal{P}_w$: Calculate for every link e the amount of shareable bandwidth, $B_z(e)$, according to Eqn. (2.1). Compute an integral minimum-cost flow f_b^k of bandwidth $B(p_w^k)$ from node s to node d in an auxiliary graph $G' = (V, E', C', \lambda)$ with available bandwidth function $B'_a : E' \to Z^+$, where

(a) $\forall\, e \in E \wedge B_a(e) > 0 \wedge e \notin p_w^i, e \in E', B'_a(e) = B_a(e)$, and $C'(e) = |V| \cdot C(e)$

(b) $\forall\, e \in E \wedge B_z(e) > 0 \wedge e \notin p_w^i, e \in E', B'_a(e) = B_z(e)$, and $C'(e) = 1$ (If link e has already been added to E' in Step 3a, this one will be parallel to the existing one; in other words, there can be two links between the same node pair in G'.)

Return NULL and undo any of the changes if f_b^k is not found; otherwise, let $\mathcal{P}_b^k$ be a set of paths constituent to the flow f_b^k (note that if two parallel links both appear in f_b^k, they should be combined in $\mathcal{P}_b^i$) and update B_a and ν_e.

4 Return $\mathcal{P}_w$ as the working VCG and $\{\mathcal{P}_b^k\}$ as the set of backup VCGs.

---

bandwidth, $B_z(e)$, as follows:

$$B_z(e) = \max_{\forall e' \in E}\{\nu_e^{e'}\} - \max_{\forall e' \in p_w^k}\{\nu_e^{e'}\} \tag{2.1}$$

For every link e whose $B_z(e) > 0$, we introduce a link e' parallel to e, i.e., links e and e' originate from the same upstream node and terminate at the same downstream node. The bandwidth of link e' will be $B_z(e)$. The cost of link e' will be far smaller than that of link e. When a minimum-cost flow algorithm is applied to a graph so modified, the algorithm will always prefer link e' to link e as long as link e' has available bandwidth because link e' has much less cost. Therefore, minimizing the cost of the flow is equivalent to maximizing backup sharing in this case.

A formal specification of PIVM is shown in Algorithm 7.1. The computational complexity of Algorithm 7.1 depends on the minimum-cost flow algorithm. If we employ the enhanced capacity-scaling algorithm, which has complexity $O((|E| \log |V|)(|E| + |V| \log |V|))$ [Ahuja et al., 1993], then the complexity of Algorithm 7.1 will be $O((u \cdot |E| \log |V|)(|E| + |V| \log |V|))$. In particular, the complexity of Step 1 is $O((|E| \log |V|)(|E| + |V| \log |V|))$; the complexity of Step 2 is $O(u \cdot |V|^2)$; the complexity of Step 3 is $O((u \cdot |E| \log |V|)(|E| + |V| \log |V|))$; and the complexity of Step 4 is $O(1)$.

We remark that PIVM is suitable for centralized implementation because: (1) PIVM needs the detailed routing information of all the existing connections, or the conflict set of all the links, to measure the backup-sharing potential for provisioning a new connection request in Step 3 of Algorithm 7.1; and (2) after a connection is provisioned, the conflict set on any link that is traversed by the backup VCGs of the connection needs to be updated. In a centralized network-management system, the detailed information of all the existing connections is available, and updating the conflict set on some links is straightforward. However, in a distributed network-management system, the detailed routing information of all the existing connections may not be available at every node. Furthermore, propagating conflict-set updates of all the links traversed by the backup VCGs incurs large signaling volume, and maintaining a consistent view of the entire network among all the nodes becomes difficult. In the following section, we propose an alternative survivable DoS approach suitable for distributed implementation.

7.3 Provisioning fast REstorable VCG (PREV)

7.3.1 Basic Idea

PREV aims at fast protection switching by judiciously sharing backup capacity such that, when a working VCG member fails, no backup configuration is needed at intermediate nodes along the backup path (but, of course, backup configuration at the source and destination nodes is necessary). As shown in [Hauser et al., 2002], to avoid backup configuration, the following constraints apply: (1) only connections having the same source-destination pair can share backup bandwidth; and (2) a backup VCG member can only be shared end-to-end. Under these constraints, PREV works as follows: (1) pre-select a backup path, p^b_{sd}, for each node pair $\langle s, d\rangle$; and (2) route each connection request such that both inter-connection sharing (with existing connections from s to d) and intra-connection sharing (by inverse multiplexing) can be maximized.

7.3.2 An Example

To elaborate on the above ideas, consider the example in Fig. 7.3. Suppose that paths p^1_{sd}, p^2_{sd}, and p^3_{sd} are link-disjoint to path p^b_{sd} and there is sufficient bandwidth available on these paths (for the purpose of illustration). Assume that path p^1_{sd} is link-disjoint to paths p^2_{sd} and p^3_{sd}. Let p^b_{sd} be the backup path for any connections from node s to node d under PREV. To provision a GbE connection from node s to node d, we can route STS-1-11v bandwidth on p^1_{sd} and STS-1-10v bandwidth on path p^2_{sd} (or p^3_{sd}). The amount of backup bandwidth that needs to be reserved on p^b_{sd} is only STS-1-11v since p^1_{sd} is link-disjoint to paths p^2_{sd} and p^3_{sd}. As a result, we can protect a GbE connection with only STS-1-11v bandwidth (about 600 Mbps) via intra-connection sharing. In general, if the

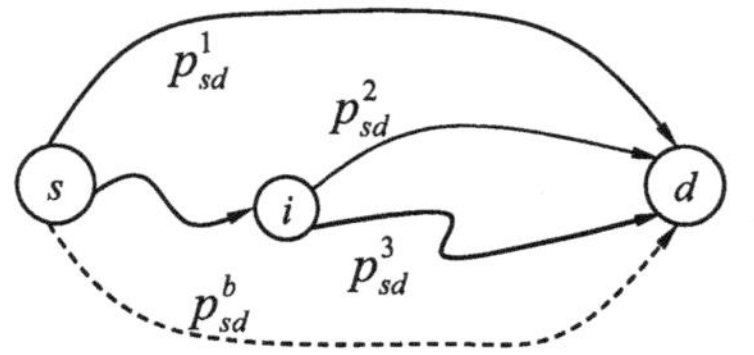

Figure 7.3. Survivable DoS approach: PREV.

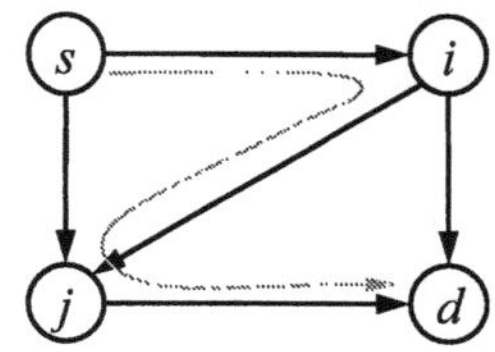

Figure 7.4. Inappropriate backup path $\langle s, i, j, d\rangle$.

number of mutually link-disjoint paths between node s and node d is larger, then the amount of backup bandwidth that needs to be reserved can be further reduced.

Later on, suppose a new connection request from node s to node d requiring bandwidth STS-1 arrives before the GbE connection leaves. The new connection can be routed on path p^2_{sd} (or p^3_{sd}) (assuming they have free bandwidth of STS-1 or higher) without reserving additional backup bandwidth on path p^b_{sd} due to inter-connection sharing.

In case of a working VCG member failure, e.g., p^1_{sd}, destination node d notifies source s via LCAS. Upon receiving the notification, source s removes the failed member (p^1_{sd}) via LCAS; configures itself to use path p^b_{sd}; and adds to the working VCG path p^b_{sd} with capacity equal to the amount of capacity previously routed on the failed working VCG member (p^1_{sd}) via LCAS. Note that there is no backup configuration at the intermediate nodes along the backup path p^b_{sd} after a failure occurs.

7.3.3 Pre-Select a Backup Path for Every Node Pair

Under PREV, a proper backup path needs to be chosen for every node pair in advance. One reasonable criterion for pre-selecting a backup path for each node pair is load balancing, i.e., to select a path p^b_{sd} for every node pair $\langle s, d\rangle$ such that the maximum number of times a link is traversed by all the backup paths $\{p^b_{sd} \mid \forall s \neq d \in V\}$ is minimized. This problem, referred to as balanced backup selection, is NP-complete for directed graph because: (1) The disjoint-path problem stated below is NP-complete for any $k \geq 2$ [Fortune et al., 1980, Robertson and Seymour, 1995]. (2) The disjoint-path problem with $k = |V| \cdot (|V| - 1)$ is a special case of the balanced backup-selection problem with the maximum number of times a link is traversed by all the backup paths being one.

Disjoint-Path Instance: A directed graph $G = (V, E)$ and distinct node pairs $\langle s_1, d_1\rangle, \cdots, \langle s_k, d_k\rangle$ ($s_i \neq d_i \in V$).

Disjoint-Path Question: Do there exist mutually link-disjoint paths, $p_1, \cdots, p_k$, in G such that p_i joins s_i and d_i ($1 \leq i \leq k$)?

Besides the load-balancing concern, a poorly-chosen backup path, p^b_{sd}, can significantly reduce the amount of working flow which is link-disjoint to p^b_{sd}. For example, Fig. 7.4 demonstrates that a poorly chosen backup path, $\langle s, i, j, d \rangle$, disconnects the source node s from the destination node d even if sufficient capacity is available on every link because the working flow needs to be link-disjoint to path $\langle s, i, j, d \rangle$. If an appropriate backup path, e.g., $\langle s, i, d \rangle$, is chosen, some link-disjoint flow can be carried on path $\langle s, j, d \rangle$. Clearly, proper approaches are needed to avoid this type of trap situation [Ou et al., 2003, Xu et al., 2003b].

Denote as $MinCut(S, D)$ a minimum cut with respect to capacity between node s and node d in $G = (V, E, C, \lambda)$ representing the given network, where $S \cup D = V$, $S \cap D = \phi$, $s \in S$, and $d \in D$ (i.e., we consider two non-overlapping partitions of the network, one partition containing the source node s and the other containing the destination node d). A "good" backup path should have the property of traversing only one link per $MinCut(S, D)$ for any $MinCut(S, D)$ to maximize the amount of working flow. (If a path traverses more than one link of a minimum-cut $MinCut(S, D)$, then the amount of flow disjoint to the path will be unnecessarily reduced at $MinCut(S, D)$ due to the equivalence of maximum flow and minimum cut.)

Let $\mathcal{P}_{sd}$ be a set of mutually link-disjoint paths between node s and node d with maximum cardinality. Clearly, any path in $\mathcal{P}_{sd}$ only traverses one link per $MinCut(S, D)$. Load-balancing criterion can then be employed to decide which path in $\mathcal{P}_{sd}$ should be selected as p^b_{sd}.

In summary, the backup path for every node pair is computed as follows:

1 For every node pair $\langle s, d \rangle$ ($s \neq d \in V$), compute $\mathcal{P}_{sd}$ by applying a maximum-flow algorithm to graph G with unit link capacity.

2 Randomly choose one path from $\mathcal{P}_{sd}$ as p^b_{sd} for every node pair $\langle s, d \rangle$.

3 Let m be the maximum number of times any link is traversed by the set of paths $\{p^b_{sd}\}$. Let L_k be the set of links traversed by k paths in $\{p^b_{sd}\}$. Find a node pair $\langle s, d \rangle$ such that: (i) p^b_{sd} traverses the most links in L_m; (ii) there exists at least one path $p \in \mathcal{P}_{sd}$ such that p does not traverse any link in L_m and L_{m-1}. Terminate if no such node pair is found; otherwise, replace p^b_{sd} by p and repeat this step.

We remark that the above procedure is guaranteed to terminate because one iteration of Step 3 reduces the size of L_m by at lease one, $|L_m| \leq |E|$, $m \leq |V| \cdot (|V| - 1)$, and m does not increase its value. The complexity of the above procedure is $O(|V|^5)$. In particular, the complexity of Step 1 is $O(|V|^5)$; the complexity of Step 2 is $O(|V|^2)$; and the complexity of Step 3 is $O(|V|^5)$. Since this procedure only needs to be run once for every topology, the high complexity may not be a problem.

7.3.4 Route Computation

Every backup path, as computed in Section 7.3.3, is associated with a conflict set to identify backup-sharing potential. The conflict set ν_{sd} for backup path p^b_{sd} can be represented as an integer set, $\{\nu^e_{sd} | \ \forall e \in E\}$, where ν^e_{sd} represents the amount of traffic that will be rerouted on backup path p^b_{sd} when link e fails. The amount of backup capacity reserved on backup path p^b_{sd} is thus $\nu^*_{sd} = \max_{\forall e \in E}\{\nu^e_{sd}\}$.

Upon the arrival of a new connection request $\langle s, d, u\rangle$, our PREV routing algorithm computes a working flow f, which is link-disjoint to the backup path p^b_{sd}, from node s to node d of bandwidth u to jointly minimize the cost of the working flow and the incremental cost on the backup path. Due to backup sharing, $(\nu^*_{sd} - \nu^e_{sd})$ units of working flow can be routed on link e without increasing the amount of backup capacity reserved on the backup path p^b_{sd}; additional amount of working flow routed on link e will cause additional amount of backup capacity to be reserved. The basic idea of our PREV routing algorithm is to route the working flow in a way such that the cost of the working flow and the incremental cost on the backup path are jointly minimized. Below, we formulate the PREV routing problem mathematically and transform the formulation to solve it efficiently.

7.3.4.1 Problem Formulation

Represent the network as a weighted, directed graph $G = (V, E, C, \lambda)$ as before. Let $B_a : E \rightarrow Z$ specify the available bandwidth on every link. Denote as C^b_{sd} the cost of the backup path p^b_{sd} and as U^b_{sd} the amount of backup bandwidth reserved on p^b_{sd} before we process the new connection request. Let $B(p^b_{sd})$ be the maximum available bandwidth along path p^b_{sd}, i.e., $B(p^b_{sd}) = \min_{\forall e \in p^b_{sd}}\{B_a(e)\}$.

Suppose that the working flow, $f : E \rightarrow Z$, is decided for the new connection request. Then, the amount of additional bandwidth to be reserved on the backup path p^b_{sd}, denoted as U_Δ, can be calculated as follows: $U_\Delta = (\max_{\forall e \in E}\{\nu^e_{sd} + f(e)\} - U^b_{sd})$. Therefore, the cost of admitting this new connection request $\langle s, d, u\rangle$ is $(\sum_{\forall e \in E} f(e) \times C(e) + U_\Delta \times C^b_{sd})$.

Since our objective is to jointly minimize the cost $(\sum_{\forall e \in E} f(e) \times C(e) + U_\Delta \times C^b_{sd})$, the working flow f can not be decided without considering the incremental cost on the backup path. However, note that U_Δ is bounded by $\min\{B(p^b_{sd}), u\})$. We can try U_Δ for every value in $[0, \min\{B(p^b_{sd}), u\}]$ to jointly minimize $(\sum_{\forall e \in E} f(e) \times C(e) + U_\Delta \times C^b_{sd})$.

The PREV routing problem can be mathematically formulated as follows. The formulation turns out to be an integer linear program (ILP). In the formulation, $B_z : E \rightarrow Z$ is a function specifying, for every link, the amount of bandwidth which can be used by the working flow and which will not cause additional bandwidth to be reserved on the backup path p^b_{sd}, i.e.,

$$B_z(e) = min\{B_a(e), U^b_{sd} - \nu^e_{sd}\}, \; \forall e \in E \tag{3.1}$$

Objective:

$$Minimize \quad (\sum_{\forall\, ij \in E} f(ij) \times C(ij)) + U_\Delta \times C^b_{sd} \tag{3.2}$$

Constraints:

$$\sum_{\forall\, j,\, ij \in E} f(ij) - \sum_{\forall\, j,\, ji \in E} f(ji) = \begin{cases} u & \text{if } i = s \\ -u & \text{if } i = d \\ 0 & \text{otherwise} \end{cases} \tag{3.3}$$

$$f(ij) \leq min\{B_a(ij), u\} \quad \forall ij \in E \tag{3.4}$$

$$f(ij) \leq B_z(ij) + U_\Delta \quad \forall ij \in E \tag{3.5}$$

$$0 \leq U_\Delta \leq min\{B(p^b_{sd}), u\} \tag{3.6}$$

$$f(ij),\, U_\Delta \; : \; int \tag{3.7}$$

Our objective function (3.2) jointly minimizes the cost of the working flow and the incremental cost on the backup path. Equation (3.3) accounts for the flow-conservation constraints for the working flow. Equation (3.4) states that the working flow can utilize no more than the amount of capacity available on any link and the amount of capacity needed by the connection request. Equation (3.5) ensures that the working flow will not cause more than U_Δ units of additional backup bandwidth to be reserved. Equation (3.6) constrains the range of the variable U_Δ.

One common approach for solving an ILP is to use a commercial solver such as CPLEX. Such an approach typically takes a long time and may not scale well given today's limited computational power. As a result, we resort to a different approach in which we only examine the crucial points in the search space without losing the optimality of the solution.

7.3.4.2 The PREV Algorithm

Our first observation is that the above ILP reduces to a standard minimum-cost flow problem if U_Δ is a constant. If U_Δ is a constant, Eqns. (3.4)-(3.5) can be transformed to constrain the available link capacity with a new function $B'_a : E \rightarrow Z$, where

$$B'_a(e) = min\{B_a(e),\; u,\; B_z(e) + U_\Delta\}. \tag{3.8}$$

The only constraint left in the ILP is Eqn. (3.3), which is exactly the specification of the minimum-cost flow problem [Ahuja et al., 1993]. Therefore, if U_Δ is a constant, the ILP can be solved by applying a minimum-cost flow algorithm to graph G with the new available link-capacity function B'_a defined in Eqn. (3.8).

Based on this observation, a straightforward approach could be to apply a minimum-cost flow algorithm for each value of $U_\Delta \in [0, min\{B(p^b_{sd}), u\}]$ and find the U_Δ corresponding to the minimum value of the objective $(\sum_{\forall e \in E} f(e) \times C(e) + U_\Delta \times C^b_{sd})$.

A more efficient approach is to apply a modified binary search on U_Δ, as discussed below. Decompose the objective function into two terms: let $c_w(U_\Delta) = \sum_{\forall e \in E} f(e) \times C(e)$ and $c_b(U_\Delta) = U_\Delta \times C^b_{sd}$ (please note that, even though there is no U_Δ in the definition of c_w, the value of U_Δ affects the working flow f, which in turn affects c_w). Clearly, $c_b(U_\Delta)$ is a linear function of U_Δ. We show that $c_w(U_\Delta)$ has the property shown in Theorem 7.1.

THEOREM 7.1

$$c_w(U^1_\Delta) - c_w(U^1_\Delta + 1) \geq c_w(U^2_\Delta) - c_w(U^2_\Delta + 1)$$
$$\forall\, 0 \leq U^1_\Delta < U^2_\Delta < min\{B(p^b_{sd}), u\} \tag{3.9}$$

Proof: *As we have shown earlier, for a fixed U_Δ, the ILP reduces to a standard minimum-cost flow problem and an optimal flow can be computed using a standard minimum-cost flow algorithm. Let f_0 be an optimal flow corresponding to a fixed U_Δ ($0 \leq U_\Delta \leq min\{B(p^b_{sd}), u\} - 2$). Let f_1 (and f_2) be an optimal flow corresponding to $U_\Delta + 1$ (and $U_\Delta + 2$).*

By Theorem 3.7 (Augmenting Cycle Theorem) in [Ahuja et al., 1993], we can convert flow f_0 to flow f_1 by augmenting along a collection of cycles C_1 in f_0's residual network with the capacity of some links increased by one. Let E_1 be the set of links whose capacity needs to be incremented. E_1 can be defined as follows:

$$E_1 = \{\forall e \in E \mid B_a(e) > (U_\Delta + B_z(e))\}$$

That is, E_1 includes those links which can carry more working flow when U_Δ increases. Using the same reasoning, we can convert flow f_0 to flow f_2 by augmenting along a collection of cycles C_2 in f_0's residual networking with the capacity of the links in E_1 increased by one and the capacity of the links in E_2 increased by one more, where E_2 is defined as:

$$E_2 = \{\forall e \in E \mid B_a(e) > (U_\Delta + B_z(e) + 1)\}$$

Clearly, any of the cycles in C_1 and C_2 has negative cost; otherwise, there is no point in augmenting the cycle. Since there is no cycle of negative cost in f_0's residual network without increasing the capacity of the links in E_1, any of the *cycles in $\mathcal{C}_1$ and $\mathcal{C}_2$ must contain at least one link from E_1 (note that $E_2 \subseteq E_1$). A possible solution for $\mathcal{C}_1$ is to choose all the cycles in $\mathcal{C}_2$ and augment by half of the capacity. Therefore, the drop in cost when we convert from flow f_0 to flow f_1 is at least half as much as that from flow f_0 to flow f_2, i.e.,*

$$c_w(U_\Delta) - c_w(U_\Delta + 1) \geq \frac{1}{2} \cdot (c_w(U_\Delta) - c_w(U_\Delta + 2))$$

or

$$c_w(U_\Delta) - c_w(U_\Delta + 1) \geq c_w(U_\Delta + 1) - c_w(U_\Delta + 2)$$

Therefore, for $0 \leq U_\Delta^1 < U_\Delta^2 < min\{B(p_{sd}^b), u\}$,

$$c_w(U_\Delta^1) - c_w(U_\Delta^1 + 1) \geq c_w(U_\Delta^1 + 1) - c_w(U_\Delta^1 + 2) \geq \cdots \geq c_w(U_\Delta^2) - c_w(U_\Delta^2 + 1)$$

■

This property provides the basis for the modified binary-search procedure. Let U_Δ^* $(0 \leq U_\Delta^* \leq min\{B(p_{sd}^b), u\})$ be the optimal value of U_Δ leading to the minimum of $c_w(U_\Delta) + c_b(U_\Delta)$. For an arbitrary U_Δ $(0 \leq U_\Delta < min\{B(p_{sd}^b), u\})$, U_Δ^* can be decided as follows.

1 If $c_w(U_\Delta) - c_w(U_\Delta + 1) < C_{sd}^b$, then $U_\Delta^* \in [0, U_\Delta]$.

2 If $c_w(U_\Delta) - c_w(U_\Delta + 1) = C_{sd}^b$, then $U_\Delta^* = U_\Delta$.

3 Otherwise, $U_\Delta^* \in [U_\Delta, min\{B(p_{sd}^b), u\}]$.

The property of c_w, as shown in Eqn. (3.9), ensures the correctness of the binary search. A formal specification of our PREV algorithm is shown in Algorithm 7.2.

The complexity of Algorithm 7.2 depends on the minimum-cost flow algorithm. If we employ the enhanced capacity-scaling algorithm, which has complexity $O((|E| \log |V|)(|E| + |V| \log |V|))$ [Ahuja et al., 1993], then the complexity of Algorithm 7.2 will be $O((\log u) \cdot (|E| \log |V|)(|E| + |V| \log |V|))$. In particular, the complexity of Step 1 is $O((|E| \log |V|)(|E| + |V| \log |V|))$; the complexity of Step 2 is $O(|E|)$; the complexity of Step 3 is $O(1)$; the complexity of Step 4 is $O((\log u) \cdot (|E| \log |V|)(|E| + |V| \log |V|))$; and the complexity of Step 5 is $O(u|V|^2)$.

In contrast to PIVM, PREV is very suitable for distributed implementation. To measure the backup-sharing potential for a new connection request, PREV only needs the conflict set of the backup path, which can be easily maintained at the source node. After a connection request is provisioned, the source node updates the conflict set of the backup path and there is no need to propagate the update. All the information which needs to be propagated is just the remaining capacity of every link that is traversed by either the working VCG or the backup path of the connection.

Algorithm 7.2 PREV.

Input: $\langle s, d, u\rangle$, $G = (V, E, C, \lambda)$, and B_a.
Output: A working VCG $\mathcal{P}_w$ and the amount of additional bandwidth U_Δ to be reserved on p^b_{sd}; otherwise NULL if no eligible solution is found.

1 Return NULL if no flow of capacity u can be found.

2 Calculate B_z according to Eqn. (3.1).

3 $U^L_\Delta \leftarrow 0$; $U^H_\Delta \leftarrow min\{B(p^b_{sd}), u\}$

4 Repeat until $U^L_\Delta = U^H_\Delta$

   (a) $U^M_\Delta \leftarrow (U^L_\Delta + U^H_\Delta)/2$

   (b) If $c_w(U^M_\Delta) - c_w(U^M_\Delta + 1) < C^b_{sd}$, then $U^H_\Delta \leftarrow U^M_\Delta$; if $c_w(U^M_\Delta) - c_w(U^M_\Delta + 1) = C^b_{sd}$, then $U^L_\Delta \leftarrow U^M_\Delta$ and $U^H_\Delta \leftarrow U^M_\Delta$; otherwise $U^L_\Delta \leftarrow U^M_\Delta$ ($c_w(U^M_\Delta)$ and $c_w(U^M_\Delta + 1)$ are computed by applying a minimum-cost flow algorithm to G with available link capacity redefined in Eqn. (3.8)).

5 Let f_w be a minimum-cost flow corresponding to U^L_Δ and $\mathcal{P}_w$ be a set of paths constituent to f_w. Return $\mathcal{P}_w$ and $U_\Delta = U^L_\Delta$.

7.4 Route Computation with Extensions to Control the Number of VCG Members

Both Algorithms 7.1 and 7.2 use a minimum-cost flow algorithm to compute VCGs. One limitation of the minimum-cost flow algorithms is that the resultant flow can have an arbitrary number of paths. In practice, we may want to limit the number of VCG members to reduce network control and management overhead, e.g., requiring that every connection request $\langle s, d, u\rangle$ can not be inversely multiplexed into more than M paths. This additional constraint significantly increases the complexity of the problem. Below, we show that it is NP-complete to compute $M = 2$ paths of combined bandwidth u and minimum cost.

Instance: A graph $G = (V, E, C, \lambda)$, available bandwidth function $B_a : E \rightarrow Z$, a connection request $\langle s, d, u\rangle$, and a positive integer w.

Question: Are there two paths from s to d of combined bandwidth no less than u and combined cost no more than w in G?

THEOREM 7.2 *The above problem, referred to as Capacitated Minimum-Cost-Multi-Path (CMCMP) problem, is NP-complete.*

Proof: Please refer to Appendix 7.A. ■

Therefore, it is NP-complete to compute up to $M (M \geq 2)$ paths of combined bandwidth u and minimum cost. We modify a minimum-cost flow algorithm to compute up to M paths of combined bandwidth u and low cost (may not be minimum cost). Our modification is based on the observation that, if the capacity of any path constituent to flow f (of capacity u) is at least $\lceil u/M \rceil$, then flow f can have no more than M paths.

Give a graph $G = (V, E, C, \lambda)$, available bandwidth function B_a, a connection request $\langle s, d, u\rangle$, and positive integer M, our modified minimum-cost flow algorithm works as follows:

1 Redefine the available bandwidth for any link $e \in E$ as $B_a^*(e) = \lfloor B_a(e)/\lceil \frac{u}{M} \rceil \rfloor$.

2 Apply a minimum-cost flow algorithm on G with available bandwidth function B_a^* to compute flow f from node s to node d of capacity $min\{u, m\}$.

3 Return failure if f is not found. Otherwise, iteratively apply a shortest-path algorithm on f to extract a set of paths $\mathcal{P}$ constituent to f. Clearly, $|\mathcal{P}| \leq M$.

4 If $u \geq M$, then: scale the capacity of any path in $\mathcal{P}$ by a factor of $\lceil \frac{u}{M} \rceil$; let p be the last path extracted by the shortest-path algorithm; and reduce the capacity of p by $(\lceil \frac{u}{M} \rceil \cdot M - u)$.

5 Return $\mathcal{P}$.

The complexity of the above algorithm is the complexity of the minimum-cost flow algorithm plus $O(|E| + M \cdot |V|^2)$. Since minimum-cost flow algorithms typically have much higher complexity than $O(|E| + M \cdot |V|^2)$ [Ahuja et al., 1993], the complexity of the above algorithm is basically the same as the complexity of the minimum-cost flow algorithm.

7.5 Performance: PIVM vs. PREV

We simulate a dynamic network environment with the assumptions that the connection-arrival processes are independent of one another at all nodes and the connection-holding time follows a negative exponential distribution. The capacity of each wavelength is OC-192 (approximately 10 Gbps), as per today's practical channel speeds. The number of connection requests follows the bandwidth (bps) distribution 50M : 100M : 150M : 600M : 1G : 2.5G : 5G : 10G = 100 : 50 : 20 : 10 : 10 : 4 : 2 : 1 (which is close to the bandwidth distribution in a practical network). Connection requests are uniformly distributed among all node pairs. An example US nationwide network topology with 16 wavelengths per fiber is shown in Fig. 7.5. Load (in Erlang) is defined as connection-arrival rate times average holding time times a connection's average bandwidth normalized in the unit of OC-192 (10 Gbps).

7.5.1 Bandwidth-Blocking Ratio

Bandwidth-blocking ratio is defined as the amount of bandwidth blocked over the amount of bandwidth offered. Please note that pure blocking probability, defined as the percentage of the *number* of connections blocked, cannot reflect the effectiveness of the algorithm as connections have different bandwidth requirements. Figure 7.6 shows that PIVM has much lower bandwidth-blocking

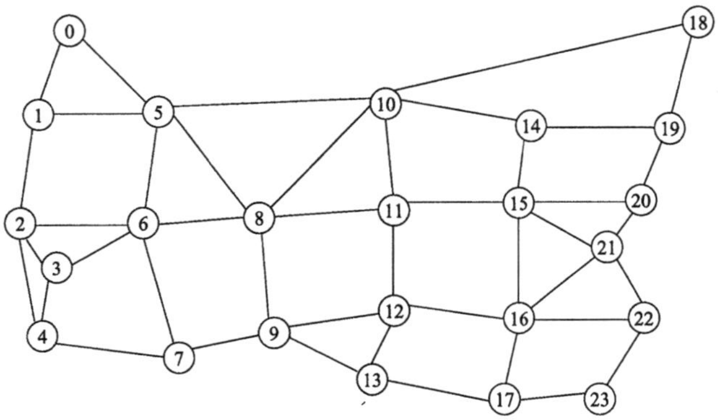

Figure 7.5. Sample US nationwide topology used in this study.

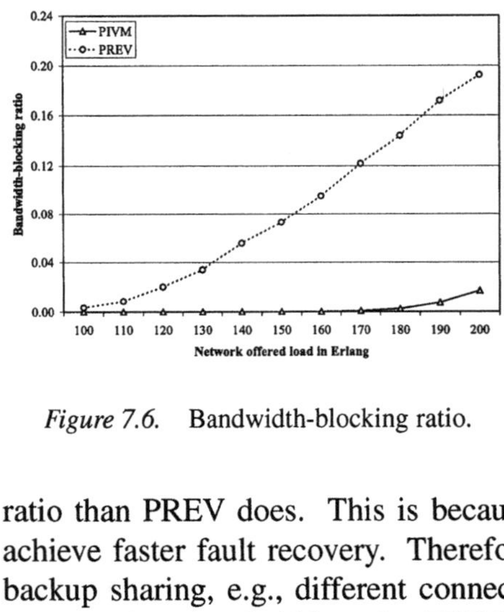

Figure 7.6. Bandwidth-blocking ratio.

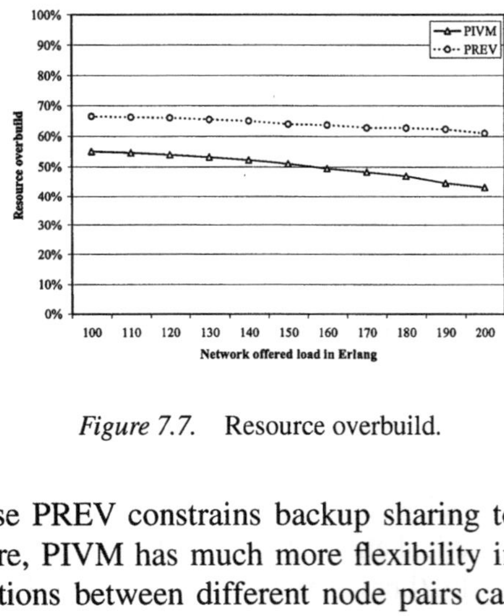

Figure 7.7. Resource overbuild.

ratio than PREV does. This is because PREV constrains backup sharing to achieve faster fault recovery. Therefore, PIVM has much more flexibility in backup sharing, e.g., different connections between different node pairs can share backup bandwidth under PIVM but not PREV, which leads to reduced bandwidth-blocking ratio.

7.5.2 Resource Overbuild

To quantify the amount of extra resources needed for providing protection as the percentage of the amount of resources required without protection, we employ a performance metric called resource overbuild [Li et al., 2002]. *Resource overbuild* is defined as the amount of backup capacity (weighted by the average hop distance of backup VCG members) over the amount of working capacity (weighted by the average hop distance of working VCG members). Figure 7.7 shows that both PREV and PIVM have much lower resource overbuild, about 60-70% for PREV and 40-60% for PIVM, than SONET/SDH's 100+% protection overhead. The resource overbuild of PIVM is lower than

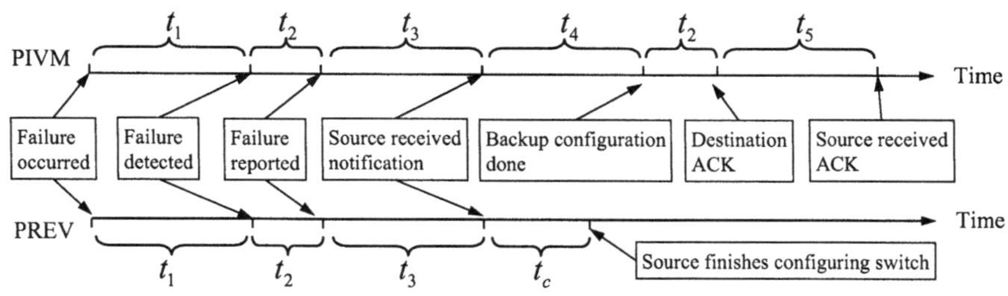

Figure 7.8. Illustration of fault-recovery time based on LCAS.

that of PREV because PREV applies constraints on backup sharing to achieve faster fault recovery.

7.5.3 Fault-Recovery Time

The reduced bandwidth-blocking ratio and increased backup sharing for PIVM comes for a price, namely, longer fault-recovery time. *Fault-recovery time* is defined as the time duration from the instant a failure occurs to the instant all the disrupted working VCG members have been successfully rerouted to their backup VCG members. The longest fault-recovery time is the maximal fault-recovery time among any possible single-link failure scenarios. Let t_p be the propagation delay of the longest path in the network and t_c be the switch-configuration time. The fault-recovery time for PIVM and PREV can be calculated as follows and is illustrated in Fig. 7.8 (the message-processing time is negligible compared to propagation delay and is not counted in the calculation).

7.5.3.1 PIVM

After a working VCG member fails, the destination node detects the failure in time t_1 (generally $t_1 \leq t_p$). The destination node reports the failure by changing the status of the failed working VCG member to FAIL. Because LCAS reports VCG member status periodically, this step may take a period of time t_2. The status of the failed member reaches the source node after time t_3 ($t_3 \leq t_p$). The source node then removes the failed VCG member, signals the nodes along the backup VCG to properly configure their switches, and changes to ADD the status of all the members in the backup VCG. The status of the backup VCG members reaches the destination node after time t_4 ($t_4 \leq t_p$). In the next reporting period (or after time t_2), the destination updates the status of the backup VCG members to OK. The new status of the backup VCG members reaches the source node in time t_5 ($t_5 \leq t_p$). At this point, if the switch at the source node has finished configuring itself, the source node can transmit payload on those backup VCG members of status OK; otherwise, the source node waits until the switch configuration is finished before it starts to transmit the payload.

In summary, the total fault-recovery time for PIVM is

$$\begin{aligned} T_{\text{PIVM}} &= t_1 + t_2 + t_3 + max\{t_c,\ t_4 + t_2 + t_5\} \\ &\leq 2 \cdot t_p + t_2 + max\{t_c,\ 2 \cdot t_p + t_2\} \end{aligned}$$

The value of t_2 can be decided as follows. LCAS reports the status of eight VCG members in a 2-ms cycle [ITU-T Rec. G.7042/Y.1305, 2001, ITU, G.707, 2002]. As we will show in Section 7.5.4 below, small VCG sizes (no more than eight) can achieve almost as good performance as large VCG sizes for typical US backbone networks, as in Fig. 7.5. As a result, LCAS can report the status of all the VCG members in one cycle for our purpose, and $t_2 = 2$ ms. For typical US backbone networks, $t_p \simeq 25$ ms. The configuration time of current switches is typically around 5 ms [Ramamurthy et al., 2003]. Therefore, the longest fault-recovery time for PIVM is about 104 ms for typical US backbone networks. Clearly, PIVM has much faster fault recovery compared to data networks such as Ethernet, whose fault recovery is on the order of seconds [Metro Ethernet Forum whitepaper, 2002].

7.5.3.2 PREV

Compared to PIVM, PREV has even faster fault recovery because, for PREV, there is no need to configure switches along the backup path except at the source node and the destination node.

After a working VCG member fails, the destination node detects the failure in time t_1 ($t_1 \leq t_p$). The destination node reports the failure in time t_2 by changing the status of the failed working VCG member to FAIL. The destination node then configures its switch. The status of the failed member reaches the source node after time t_3 ($t_3 \leq t_p$). The source node then removes the failed VCG member, changes to ADD the status of the backup path with appropriate capacity, and configures its switch. Once the switch configuration is done at the source node. The source node can start to transmit payload on the backup path. In summary, the total fault-recovery time for PREV is

$$\begin{aligned} T_{\text{PREV}} &= t_1 + t_2 + t_3 + t_c \\ &\leq t_p + t_2 + t_p + t_c \end{aligned}$$

Therefore, under PREV, the longest fault-recovery time for typical US backbone networks is 57 ms ($t_p \simeq 25$ ms, $t_2 = 2$ ms, and $t_c = 5$ ms), which is comparable to SONET 50-ms protection.

Please note that we consider the longest fault-recovery time above. Since a typical connection traverses less than half of the network diameter and signal-propagation delay is the dominant component of fault-recovery time, the av-

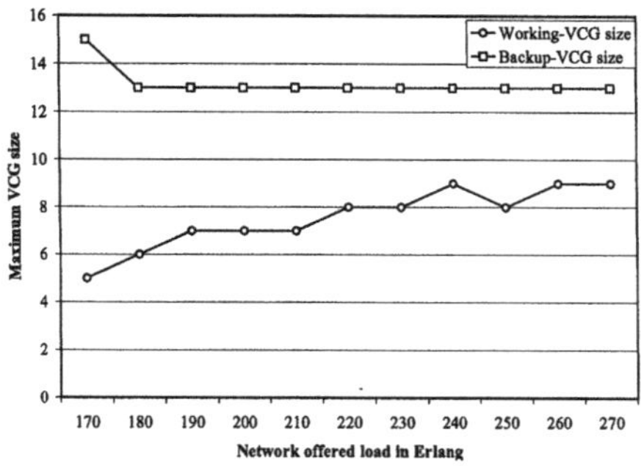

(a) PIVM.

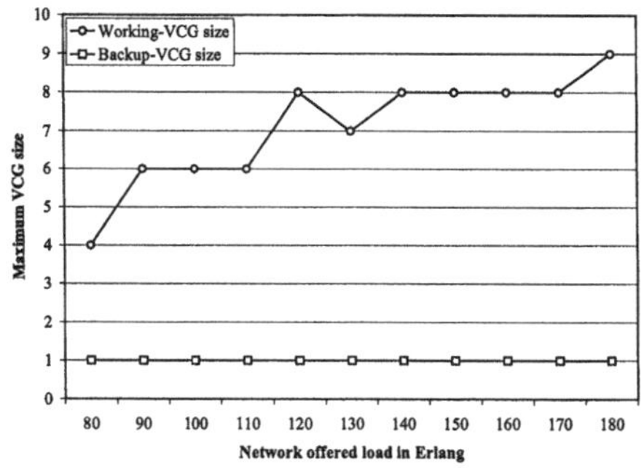

(b) PREV.

Figure 7.9. Maximum VCG size ($M = +\infty$ for both working and backup VCGs).

erage fault-recovery time will be much shorter than the longest fault-recovery time.

It can also be observed from the above discussion that PREV has very little signaling overhead after a failure occurs. As a result, PREV requires less control bandwidth compared to PIVM.

7.5.4 Impact of VCG Size

So far, there was no constraint on VCG size. Figure 7.9 shows that both the working-VCG size and the backup-VCG size for PIVM and the working-VCG size for PREV can be quite large. While large VCG size may improve network performance in terms of bandwidth-blocking ratio, large VCG size may not be desirable in practice because it increases implementation complexity and NC&M overhead. A natural question is: what is the "optimal" VCG size to strike a good balance between performance and complexity?

Figure 7.10, which resulted from our heuristic on controlling the number of VCG members described in Section 7.4, shows that small VCG sizes can achieve almost as good performance as large VCG sizes. As shown in Fig. 7.10(a), for PIVM, the bandwidth-blocking ratio for the case in which working VCGs can be inversely multiplexed on up to four paths and backup VCGs can be inversely multiplexed on up to two paths is close to the case in which both working and backup VCGs can be inversely multiplexed on an unrestricted number of paths. For PREV, as shown in Fig. 7.10(b), the bandwidth-blocking ratio for the case in which working VCGs can be inversely multiplexed on up to two paths is almost the same as the case in which working VCGs can be inversely multiplexed on an unrestricted number of paths. The reason that $M = 2$ is a good choice for PREV is that the average number of mutually link-disjoint paths for the topology shown in Fig. 7.5 is about three. Since the backup path for every node pair is chosen from a set of mutually link-disjoint paths of maximum cardinality,

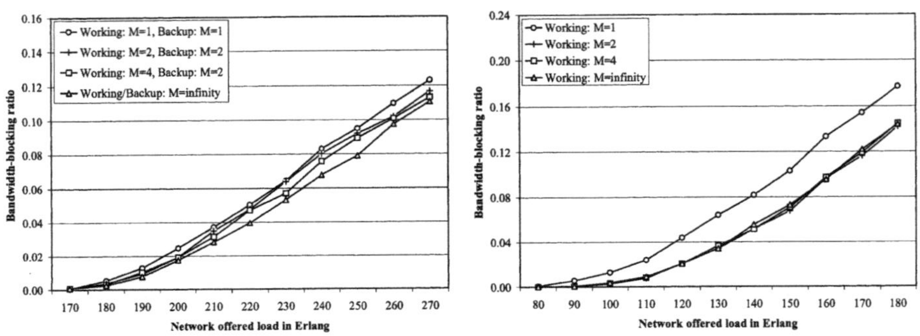

(a) PIVM. (b) PREV.

Figure 7.10. Impact of VCG size on bandwidth-blocking ratio.

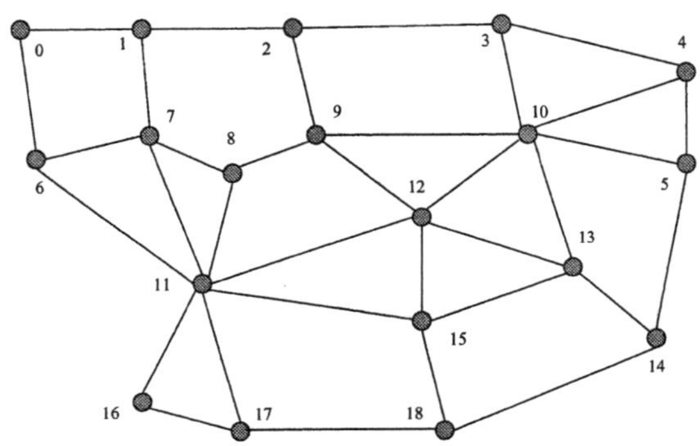

Figure 7.11. Another sample topology.

the remaining two mutually link-disjoint paths, which are link-disjoint to the backup path, will be chosen to carry the working traffic because this setting leads to maximum backup sharing (please refer to Algorithm 7.2). The fundamental law here is the maximum-flow minimum-cut associated with the topology.

For a different topology shown in Fig. 7.11, Fig. 7.12 demonstrates the same observation, i.e., small VCG sizes can achieve almost as good performance as large VCG sizes (the specific "optimal" values of M might be topology dependent).

7.6 Conclusion

We investigated the survivability of data over SONET/SDH, which is gaining increasing attention because, by such a mechanism, network operators can provide integrated data and voice services over their optical transport network to generate new revenue. We proposed two approaches—PIVM, which is suitable for centralized implementation, and PREV, which is suitable for dis-

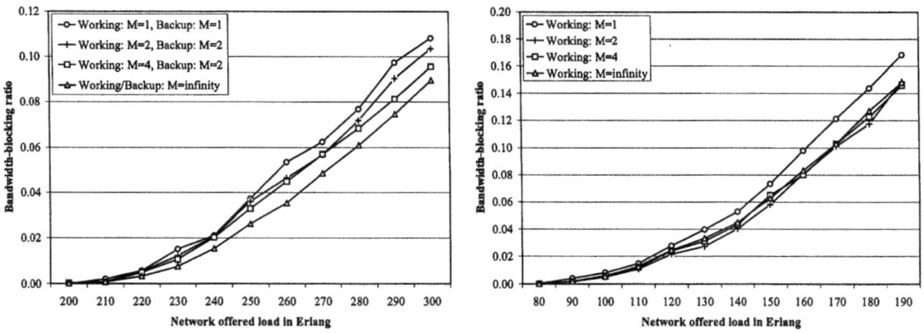

(a) PIVM. (b) PREV.

Figure 7.12. Impact of VCG size on bandwidth-blocking ratio for the topology shown in Fig. 7.11.

tributed implementation—for provisioning survivable DoS connections. Our approaches exploit the tradeoff between resource overbuild and fault-recovery time while utilizing the inverse-multiplexing capability of virtual concatenation to increase backup sharing. Our results demonstrated that PIVM achieves low resource overbuild and much faster fault recovery than that of data networks, and PREV achieves fast fault recovery comparable to SONET 50-ms protection (for typical US backbone networks) while still achieving modest backup sharing.

We further investigated the impact of VCG size on network performance. We proved that it is NP-complete to compute a minimum-cost VCG having two members of combined bandwidth no less than a given value. Results from our effective heuristic showed that, by inversely multiplexing a connection into a few paths (depending on topology) for both working and backup VCGs, the network blocking performance is almost as good as the case in which a connection can be inversely multiplexed on an unlimited number of paths.

APPENDIX 7.A: NP-Completeness of the CMCMP Problem

Basic idea: We reduce 3SAT, which is known to be NP-complete [Garey and Johnson, 1979], to the CMCMP problem. The 3SAT problem is formally stated as follows.

3SAT Instance: Collection $F = \{D_1, D_2, \ldots, D_m\}$ of clauses on a finite set Q of variables such that $|D_i| = 3$ for $1 \leq i \leq m$, where clause D_i is the Boolean "or" of three literals (a literal is either a variable or the Boolean "not" of a variable) and is satisfied by a truth assignment if and only if at least one of the three literals is true. $|F| = m$, where m is the number of clauses; $|Q| = n$, where n is the number of variables.

3SAT Question: Is there a truth assignment for Q that satisfies all the clauses in F?

The basic idea of our proof is as follows: construct, for an arbitrary instance of 3SAT, a graph such that there is a path p_1 corresponding to the false assignment of all the variables and a path p_2 corresponding to the truth assignment of all the clauses. The cost and capacity of every link are judiciously constructed in a way such that paths p_1 and p_2 are of combined *capacity no less than u and combined cost no more than w if and only if the given instance of 3SAT is satisfied.* The formal proof follows.

Proof of Theorem 7.2: CMCMP $\in$ NP since a non-deterministic algorithm can guess two paths, p_1 and p_2, and check in polynomial time if these two paths are of combined capacity no less than u and combined cost no more than w.

Given an arbitrary instance of 3SAT $F = \{D_1, D_2, \ldots, D_m\}$ and $Q = \{v_1, v_2, \ldots, v_n\}$, we construct in polynomial time an instance of CMCMP $G = (V, E, C, \lambda)$, B_a, s, and d, where $\lambda(e) = u$ for all $e \in E$. In the following construction, we define the set of nodes V, the set of links E, the cost function C, and the available bandwidth function B_a, respectively.

The set of nodes V consists of a source node s and a destination node d. The other nodes in V can be divided into two groups. The first group is related to the variables in Q and consists of the nodes $a_k, b_j^i, \overline{b}_j^i, c_j^i$, and $\overline{c}_j^i$, where $0 \leq k \leq n, 1 \leq i \leq m, 1 \leq j \leq n$. The second group of nodes is related to the clauses in F and consists of the nodes x_i and y_i for $1 \leq i \leq m$.

The set of links (unidirectional fibers) E can also be divided into two groups. In the reduction, the first group of links will be used for assigning values to the variables in Q; the second group of links will be used for evaluating the Boolean value of the clauses. The two groups of links are as follows.

- Group 1:
  (a) A link from node b_j^i to node c_j^i, $1 \leq i \leq m$ and $1 \leq j \leq n$.
  (b) A link from node $\overline{b}_j^i$ to node $\overline{c}_j^i$, $1 \leq i \leq m$ and $1 \leq j \leq n$.
  (c) A link from node c_j^i to node b_j^{i+1}, $1 \leq i \leq m-1$ and $1 \leq j \leq n$.
  (d) A link from node $\overline{c}_j^i$ to node $\overline{b}_j^{i+1}$, $1 \leq i \leq m-1$ and $1 \leq j \leq n$.
  (e) A link from node a_j to node b_{j+1}^1, and a link from node a_j to node $\overline{b}_{j+1}^1$, $0 \leq j \leq n-1$.
  (f) A link from node c_j^m to node a_j, and a link from node $\overline{c}_j^m$ to node a_j, $1 \leq j \leq n$.
- Group 2:
  (a) A link from node x_i to b_j^i, and a link from node c_j^i to node y_i, if and only if the variable v_j is in clause D_i.
  (b) A link from node x_i to $\overline{b}_j^i$, and a link from node $\overline{c}_j^i$ to node y_i, if and only if $\overline{v}_j$ is in clause D_i.
  (c) A link from node y_i to node x_{i+1}, $1 \leq i \leq m-1$.

In addition to the above links, we also have a link from node s to node a_0, a link from node a_n to node d, a link from node s to node x_1, and a link from node y_m to node d.

The link-cost function, $C : E \rightarrow Z$, is defined as follows:

$$C(e) = \begin{cases} \lceil \frac{w}{2} \rceil - m \cdot n & \text{if } e = \langle s, a_0 \rangle \\ \lfloor \frac{w}{2} \rfloor & \text{if } e = \langle s, x_1 \rangle \\ 1 & \text{if } e = \langle c_j^i, b_j^{i+1} \rangle, \text{ or } e = \langle \overline{c}_j^i, \overline{b}_j^{i+1} \rangle, \text{ or } e = \langle c_j^m, a_j \rangle, \text{ or } e = \langle \overline{c}_j^m, a_j \rangle \\ & (1 \leq i \leq m-1 \text{ and } 1 \leq j \leq n) \\ 0 & \text{otherwise} \end{cases}$$

The available bandwidth function, $B_a : E \rightarrow Z$, is defined as follows:

$$B_a(e) = \begin{cases} 1 & \text{if } e \text{ originates or terminates at node } x_i \text{ or node } y_i \ (1 \leq i \leq m) \\ u - 1 & \text{otherwise} \end{cases}$$

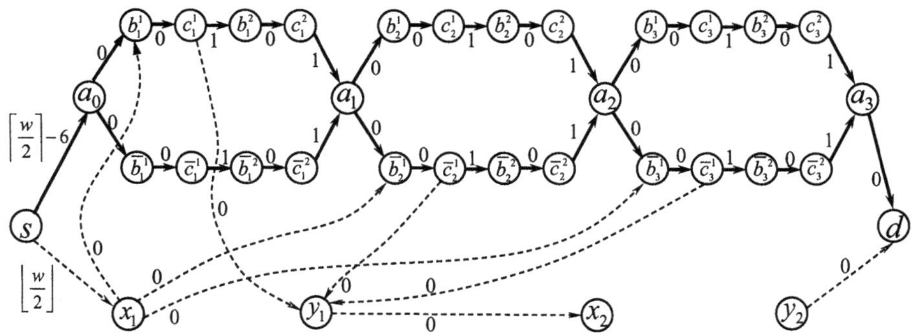

Figure 7.A.1. An illustrative construction for a 3SAT instance $F = \{D_1, D_2\}, D_1 = v_1 \vee \bar{v}_2 \vee \bar{v}_3, D_2 = \bar{v}_1 \vee v_2 \vee v_3$, and $Q = \{v_1, v_2, v_3\}$ (the part corresponding to clause D_2 is not shown). The number on every link is the cost. The dashed lines have one unit of available capacity, and the solid lines have $u - 1$ units of available capacity.

It is easy to see that the construction can be done in polynomial time. An illustrative construction (the construction for the second clause is similar to the first one and is not shown) for a 3SAT instance $F = \{D_1, D_2\}, D_1 = v_1 \vee \bar{v}_2 \vee \bar{v}_3, D_2 = \bar{v}_1 \vee v_2 \vee v_3$, and $Q = \{v_1, v_2, v_3\}$ is shown in Fig. 7.A.1.

We now show that, if F is satisfiable, then from node s to node d in graph G there exist two paths of combined cost w and combined bandwidth u. Let $v_1 = z_1, v_2 = z_2, \ldots, v_n = z_n$ be an assignment that satisfies F, where $z_j \in \{0, 1\}$ for $1 \leq j \leq n$. The two paths can be routed as follows. The first path, p_1, is routed via the nodes b_j^i and c_j^i ($1 \leq i \leq m$ and $1 \leq j \leq n$) if and only if $z_j = 0$; otherwise, p_1 is routed via nodes $\bar{b}_j^i$ and $\bar{c}_j^i$ (p_1 will also need to traverse all the a_k nodes for $0 \leq k \leq n$). The second path, p_2, is routed via x_i, y_i, and other nodes defined as follows. By the construction, the path from node x_i to node y_i corresponds to clause $D_i, |D_i| = 3$. Without loss of generality, let $D_i = v_f \vee \bar{v}_g \vee v_h$, where f, g, and h are distinct integers between 1 and n. Then, there will be three paths from node x_i to node y_i that go through nodes $b_f^i, \bar{b}_g^i$, and b_h^i, respectively. Since $v_1 = z_1, v_2 = z_2, \ldots, v_n = z_n$ is an assignment that satisfies F, either $z_f = 1$, or $z_g = 0$, or $z_h = 1$. If $z_f = 1$, then path p_2 traverses links $\langle b_f^i, c_f^i \rangle$ and $\langle c_f^i, y_i \rangle$; if $z_g = 0$, then path p_2 traverses links $\langle \bar{b}_g^i, \bar{c}_g^i \rangle$ and $\langle \bar{c}_g^i, y_i \rangle$; if $z_h = 1$, then path p_2 traverses links $\langle b_h^i, c_h^i \rangle$ and $\langle c_h^i, y_i \rangle$; if more than one condition is true, then randomly pick one.

Paths p_1 and p_2 so selected have combined bandwidth u and combined cost w due to the following facts:

1 The capacity of path p_2 is unity.

2 The capacity of path p_1 is $u - 1$ because p_1 is link-disjoint to p_2, which can be shown as follows. Path p_1 only traverses links in Group 1, and the Group-1 links which p_1 traverses correspond to the literals of value 0. Path p_2 traverses links in both Group 1 and Group 2, and the Group-1 links which p_2 traverses correspond to the literals of value 1. By the construction, a Group-1 link corresponding to a literal of value 0 is disjoint to a Group-1 link corresponding to a literal of value 1; and a Group-1 link is disjoint to a Group-2 link. Therefore, paths p_1 and p_2 are link-disjoint. Because all the links path p_1 traverses have capacity $u - 1$, the capacity of p_1 is $u - 1$.

3 The cost of path which p_2 is $\lfloor \frac{w}{2} \rfloor$ because all the Group-1 links p_2 traverses have the cost of zero.

4 The cost of path p_1 is $\lceil \frac{w}{2} \rceil$ because: (i) p_1 traverses link $\langle s, a_0 \rangle$, which has cost $\lceil \frac{w}{2} \rceil - m \cdot n$. (ii) p_1 traverses either links $\langle c_j^i, b_j^{i+1} \rangle$ and $\langle c_j^m, a_j \rangle$ or links $\langle \bar{c}_j^i, \bar{b}_j^{i+1} \rangle$ and $\langle \bar{c}_j^m, a_j \rangle$ ($1 \leq i \leq m-1$ and $1 \leq j \leq n$), all of which have unity cost. The total cost of the links in this category is $(m-1) \cdot n + n = m \cdot n$. (iii) All the other links which p_1 traverses are of zero cost.

Thus, we have shown that, if F is satisfiable, then we can find two paths of combined bandwidth u and combined cost w.

We now show that, if there are two paths p_1 and p_2 from node s to node d of combined bandwidth u and combined cost w in graph G, then F is satisfiable due to the following facts.

1 Since the amount of outgoing flow at node s is u, both links $\langle s, a_0 \rangle$ and $\langle s, x_1 \rangle$ will be used by paths p_1 and p_2. Without loss of generality, suppose that path p_1 traverses link $\langle s, a_0 \rangle$, and path p_2 traverses link $\langle s, x_1 \rangle$.

2 The capacity of path p_1 is $u-1$ and the capacity of path p_2 is unity because: (i) the combined capacity of paths p_1 and p_2 is u; (ii) path p_1 traverses link $\langle s, a_0 \rangle$, which has available capacity $u-1$; (iii) path p_2 traverses link $\langle s, x_1 \rangle$, which has unity available capacity.

3 Path p_1 can only traverse links in Group 1; otherwise, the capacity of path p_1 will be smaller than $u-1$ and the combined capacity of the two paths will be smaller than u.

4 Paths p_1 and p_2 are link-disjoint because the capacity of any of the Group-1 links is $u-1$.

5 Path p_1 traverses either links $\langle c_j^i, b_j^{i+1} \rangle$ and $\langle c_j^m, a_j \rangle$ or links $\langle \bar{c}_j^i, \bar{b}_j^{i+1} \rangle$ and $\langle \bar{c}_j^m, a_j \rangle$ ($1 \leq i \leq m-1$ and $1 \leq j \leq n$), all of which have unity cost. The total cost of the links in this category is $(m-1) \cdot n + n = m \cdot n$. Therefore, the cost of path p_1 is $\lceil \frac{w}{2} \rceil$.

6 The cost of path p_2 can be no more than $\lfloor \frac{w}{2} \rfloor$ because the combined cost of paths p_1 and p_2 is w. Path p_2 can not traverse any Group-1 links of non-zero cost because p_2 traverses link $\langle s, x_1 \rangle$ of cost $\lfloor \frac{w}{2} \rfloor$.

7 We represent the route for path p_1 by a n-bit binary number, $z_1 z_2 \ldots z_n$. Path p_1 is routed as follows: if bit z_j is 1, then p_1 is routed over the nodes $\bar{b}_j^i$; otherwise, it is routed over the nodes b_j^i. Note that there is a one-to-one mapping between an n-bit binary number and a path using only Group-1 links in the network. Let $z_1 z_2 \ldots z_n$ be the n-bit number that corresponds to path p_1. Then, there is a path from node x_i to node y_i that does not use the links traversed by p_1, if and only if clause D_i is true under the following assignment: $v_j = \bar{z}_j$, where $1 \leq j \leq n$. The reason is as follows. Without loss of generality, let $D_i = v_f \vee \overline{v}_g \vee v_h$, where f, g, and h are distinct integers between 1 and n. There are three paths from node x_i to node y_i that traverse nodes b_f^i, $\bar{b}_g^i$, and b_h^i (note that path p_2 cannot traverse any Group-1 links of non-zero cost). Path p_1 will traverse the nodes b_f^i, $\bar{b}_g^i$, and b_h^i, if and only if $z_f = 0$, $z_g = 1$, and $z_h = 0$, respectively. Thus, there exists a path from node x_i to y_i if and only if $z_f = 1$, or $z_g = 0$, or $z_h = 1$, which is exactly the condition under which D_i will be true.

Since paths p_1 and p_2 have combined capacity u and combined cost w, D_i is true for all $1 \leq i \leq m$ due to the above facts. Consequently, F is satisfiable under the following assignment: $v_j = z_j, 1 \leq j \leq n$, where $z_1 z_2 \ldots z_n$ is the n-bit number corresponding to path p_1.

This concludes our proof that CMCMP is NP-complete.

References

[Ahuja et al., 1993] Ahuja, R. K., Magnanti, T. L., and Orlin, J. B. (1993). *Network Flows: Theory, Algorithms, and Applications*. Prentice-Hall.

[Akyamac et al., 2002] Akyamac, A. A., Sengupta, S., Labourdette, Jean-Francois, Chaudhuri, S., and French, S. (2002). Reliability in single domain vs. multi domain optical mesh networks. In *Proc. National Fiber Optic Engineers Conference*.

[Anand et al., 2002] Anand, Vishal, Chauhan, Sunit, and Qiao, Chunming (Dept. of CSE, State University of New York at Buffalo, Tech. Report 2002-01, Jan. 2002). Sub-path protection: A new framework for optical layer survivability and its quantitative evaluation.

[ANSI T1X1.5, 2001-062, 2001] ANSI T1X1.5, 2001-062 (2001). Synchronous optical networks (SONET).

[Assi et al., 2003] Assi, C., Ye, Y., Shami, A., Dixit, S., and Ali, M. (2003). Efficient path selection and fast restoration algorithms for shared restorable optical networks. In *Proc. IEEE ICC*, pages 1412–1416.

[Banerjee and Mukherjee, 1996] Banerjee, D. and Mukherjee, B. (1996). A practical approach for routing and wavelength assignment in large wavelength-routed optical networks. *IEEE J. Selected Areas in Communications*, 14(5):903–908.

[Bhandari, 1999] Bhandari, Ramesh (1999). *Survivable Networks: Algorithms for Diverse Routing*. Kluwer Academic Publishers.

[Bouillet et al., 2002a] Bouillet, E., Labourdette, Jean-Francois, Ellinas, G., Ramamurthy, R., and Chaudhuri, S. (2002a). Stochastic approaches to compute shared mesh restored lightpaths in optical network architectures. In *Proc. IEEE INFOCOM*, pages 801–807.

[Bouillet et al., 2002b] Bouillet, E., Labourdette, Jean-Francois, Ramamurthy, R., and Chaudhuri, Sid (2002b). Enhanced algorithm cost model to control tradeoffs in provisioning shared mesh restored lightpaths. In *Proc. OFC*, page ThW2.

[Chen and Banerjee, 1996] Chen, Chien and Banerjee, S. (1996). A new model for optimal routing and wavelength assignment in wavelength division multiplexed optical networks. In *Proc. IEEE INFOCOM*, volume 1, pages 164–171.

[Chiu and Strand, 2003] Chiu, A. and Strand, J. (2003). Control plane considerations for all-optical and multi-domain optical networks and their status in OIF and IETF. *SPIE Optical Networks Magazine*, 4:26–35.

[Chlamtac et al., 1996] Chlamtac, I., Farago, A., and Zhang, T. (1996). Lightpath (wavelength) routing in large WDM networks. *IEEE J. Selected Areas in Communications*, 14(5):909–913.

[Chlamtac et al., 1992] Chlamtac, I., Ganz, A., and Karmi, G. (1992). Lightpath communications: an approach to high bandwidth optical WAN's. *IEEE Trans. Communications*, 40(7):1171–1182.

[Cisco white paper, phtm] Cisco white paper (1999. Available at http://www.cisco.com/warp/public/cc/so/neso/olso/wartso/opdn_wp.htm). Scaling optical data networks with cisco wavelength routing.

[Cormen et al., 1990] Cormen, T. H., Leiserson, C. E., and Rivest, R. L. (1990). *Introduction to Algorithms*. McGraw-Hill, New York City, NY.

[Crochat et al., 2000] Crochat, O., Le Boudec, J.-Y., and Gerstel, O. (2000). Protection interoperability for WDM optical networks. *IEEE/ACM Trans. Networking*, 8(3):384–395.

[Datta et al., 2001] Datta, S., Sengupta, S., Biswas, S., and Datta, S. (2001). Efficient channel reservation for backup paths in optical mesh networks. In *Proc. IEEE Globecom*, pages 2104–2108.

[Doshi et al., 1999] Doshi, Bharat T., Dravida, Subrahmanyam, Harshavardhana, P., Hauser, Oded, and Wang, Yufei (1999). Optical network design and restoration. *Bell Labs Technical Journal*, 4(1):58–84.

[Doverspike and Yates, 2001] Doverspike, R. and Yates, J. (2001). Challenges for MPLS in optical network restoration. *IEEE Communication Magazine*, 39(2):89–96.

[Dunn et al., 1994] Dunn, D.A., Grover, W.D., and MacGregor, M.H. (1994). Comparison of k-shortest paths and maximum flow routing for network facility restoration. *IEEE J. Selected Areas in Communications*, 12(1):88–99.

[Elie-Dit-Cosaque et al., 2002] Elie-Dit-Cosaque, David, Ali, Maher, and Tancevski, Lubo (2002). Informed dynamic shared path protection. In *Proc. OFC*, page ThO4.

[Ellinas et al., 2003] Ellinas, G., Bouillet, E., Ramamurthy, R., Labourdette, J., Chaudhuri, S., and Bala, K. (2003). Routing and restoration architectures in mesh optical networks. *SPIE Optical Networks Magazine*, 4(1):91–106.

[Ellinas et al., 2001] Ellinas, G., Hailemariam, A.G., and Stern, T. E. (2001). Protection cycles in mesh WDM networks. *IEEE J. Selected Areas in Communications*, 18(10):1924–1937.

[Foisel, 2001] Foisel, H.M. (OIF 2001.349, July 2001). NNI: Dual node interconnection.

[Fortune et al., 1980] Fortune, S., Hopcroft, J. E., and Wyllie, J. (1980). The directed subgraph homeomorphism problem. *J. Theoretical Computer Science*, 10(2):111–121.

[Frieze, 2001] Frieze, Alan M. (2001). Edge-disjoint paths in expander graphs. *Society for Industrial and Applied Mathematics Journal on Computing*, 30(6):1790–1801.

[Fumagalli et al., 2003] Fumagalli, A., Cerutti, I., and Tacca, M. (2003). Optimal design of survivable mesh networks based on line switched WDM self-healing rings. *IEEE/ACM Trans. Networking*, 11(3):501–512.

[Fumagalli et al., 1999] Fumagalli, A., Cerutti, I., Tacca, M., Masetti, F., Jagannathan, R., and Alagar, S. (1999). Survivable networks based on optimal routing and WDM self-healing rings. In *Proc. IEEE INFOCOM*, volume 2, pages 726–733.

[Fumagalli and Valcarenghi, 2000] Fumagalli, A. and Valcarenghi, L. (2000). IP restoration vs. WDM protection: is there an optimal choice. *IEEE Network*, 14(6):34–41.

[Gardner et al., 1994] Gardner, L.M., Heydari, M., Shah, J., Sudborough, I.H., Tollis, I.G., and Xia, C. (1994). Techniques for finding ring covers in survivable networks. In *Proc. IEEE Globecom*, volume 3, pages 1862–1866.

[Garey and Johnson, 1979] Garey, M. R. and Johnson, D. S. (1979). *Computers and Intractability: A Guide to the Theory of NP-Completeness*. W. H. Freeman Company, New York.

[Gerstel and Ramaswami, 2000a] Gerstel, O. and Ramaswami, R. (2000a). Optical layer survivability–an implementation perspective. *IEEE J. Selected Areas in Communications*, 18(10):1885–1899.

[Gerstel and Ramaswami, 2000b] Gerstel, O. and Ramaswami, R. (2000b). Optical layer survivability: a services perspective. *IEEE Communication Magazine*, 38(3):104–113.

[Gerstel et al., 2000] Gerstel, O., Ramaswami, R., and Sasaki, G. H. (2000). Cost-effective traffic grooming in WDM rings. *IEEE/ACM Trans. Networking*, 8:618–630.

[Goyal et al., 2002] Goyal, Mukul, Li, Guangzhi, and Yates, Jennifer (2002). Shared mesh restoration: A simulation study. In *Proc. OFC*, pages 489–490.

[Grover, 1999] Grover, W.D. (1999). High availability path design in ring-based optical networks. *IEEE/ACM Trans. Networking*, 7(4):558–574.

[Grover and Stamatelakis, 1998] Grover, W.D. and Stamatelakis, D. (1998). Cycle-oriented distributed preconfiguration: ring-like speed with mesh-like capacity for self-planning network restoration. In *Proc. IEEE ICC*, volume 1, pages 537–543.

[Gummadi et al., 2003] Gummadi, K.P., Pradeep, M.J., and Murthy, C.S.R. (2003). An efficient primary-segmented backup scheme for dependable real-time communication in multihop networks. *IEEE/ACM Trans. Networking*, 11(1):81–94.

[Hac, 1994] Hac, Anna (1994). Improving reliability through architecture partitioning in telecommunication networks. *IEEE J. Selected Areas in Communications*, 12(4):193–204.

[Hauser et al., 2002] Hauser, Oded, Kodialam, M., and Lakshman, T. V. (2002). Capacity design of fast path restorable optical networks. In *Proc. IEEE INFOCOM*, volume 2, pages 817–826.

[Hernandez-Valencia et al., 2002] Hernandez-Valencia, E., Scholten, M., and Zhu, Zhenyu (2002). The generic framing procedure (GFP): an overview. *IEEE Communication Magazine*, 40(5):63–71.

[Ho and Mouftah, 2002] Ho, Pin-Han and Mouftah, H.T. (2002). A framework for service-guaranteed shared protection in WDM mesh networks. *IEEE Communication Magazine*, 40:97–103.

[Iness and Mukherjee, 1999] Iness, J. and Mukherjee, B. (1999). Sparse wavelength conversion in wavelength-routed WDM networks. *J. Photonic Network Communications*, 1(3):183–205.

[Iraschko et al., 1998] Iraschko, R.R., MacGregor, M.H., and Grover, W.D. (1998). Optimal capacity placement for path restoration in STM or ATM mesh-survivable networks. *IEEE/ACM Trans. Networking*, 6(3):325–336.

[ITU, G.707, 2002] ITU, G.707 (2002). Network node interface for the synchronous digital hierarchy (SDH).

[ITU-T Rec. G.7042/Y.1305, 2001] ITU-T Rec. G.7042/Y.1305 (2001). Link capacity adjustment scheme (LCAS) for virtual concatenated signals.

[Kar et al., 2002] Kar, Koushik, Kodialam, M., and Lakshman, T. V. (2002). Routing restorable bandwidth guaranteed connections using maximum 2-route flows. In *Proc. IEEE INFOCOM*, pages 113–121.

[Kawamura et al., 1994] Kawamura, R., Sato, K.-I., and Tokizawa, I. (1994). Self-healing ATM networks based on virtual path concept. *IEEE J. Selected Areas in Communications*, 12(1):120–127.

[Kodialam and Lakshman, 2000] Kodialam, M. and Lakshman, T. V. (2000). Dynamic routing of bandwidth guaranteed tunnels with restoration. In *Proc. IEEE INFOCOM*, volume 2, pages 902–911.

[Kodialam and Lakshman, 2001] Kodialam, M. and Lakshman, T. V. (2001). Dynamic routing of locally restorable bandwidth guaranteed tunnels using aggregated link usage information. In *Proc. IEEE INFOCOM*, volume 1, pages 376–385.

[Kodialam and Lakshman, 2003] Kodialam, M. and Lakshman, T. V. (2003). Dynamic routing of restorable bandwidth-guaranteed tunnels using aggregated network resource usage information. *IEEE/ACM Trans. Networking*, 11(3):399–410.

[Koo and Subramaniam, 2002] Koo, S. and Subramaniam, S. (2002). Trade-offs between speed, capacity, and restorability in optical mesh network restoration. In *Proc. OFC*, pages 487–489.

[Kuipers et al., 2002] Kuipers, F., Mieghem, P. V., Korkmaz, T., and Krunz, M. (2002). An overview of constraint-based path selection algorithms for qos routing. *IEEE Communication Magazine*, 40:50–55.

[Lardies et al., 2001] Lardies, Ana, Gupta, Rakesh, and Patterson, Raymond A. (2001). Traffic grooming in a multi-layer network. *SPIE Optical Networks Magazine*, 2:91–99.

[Li et al., 2002] Li, Guangzhi, Wang, Dongmei, Kalmanek, Charles, and Doverspike, Robert (2002). Efficient distributed path selection for shared restoration connections. In *Proc. IEEE INFOCOM*, pages 140–149.

[Li et al., 2003] Li, J., Park, H., and Lee, H. (2003). Shared sub-path protection with overlapped protection areas in WDM networks. In *Proc. OFC*, page FQ7.

[Liu and Tipper, 2001] Liu, Y. and Tipper, D. (2001). Successive survivable routing for node failures. In *Proc. IEEE Globecom*, volume 4, pages 2093–2097.

[Liu et al., 2001] Liu, Yu, Tipper, D., and Siripongwutikorn, P. (2001). Approximating optimal spare capacity allocation by successive survivable routing. In *Proc. IEEE INFOCOM*, volume 2, pages 699–708.

[Lumetta et al., 2000] Lumetta, S. S., Medard, M., and Tseng, Yung-Ching (2000). Capacity versus robustness: a tradeoff for link restoration in mesh networks. *IEEE/OSA J. Lightwave Technology*, 18(12):1765–1775.

[MacDonald et al., 2000] MacDonald, R., Chen, L.-P., Shi, C.-X., and Faer, B. (2000). Requirements of optical layer network restoration. In *Proc. OFC*, pages 68–70.

[Matula et al., 1972] Matula, D. W., Marble, G., and Isaacson, J. D. (1972). Graph coloring algorithms. In Read, R. C., editor, *Graph Theory and Computing*, chapter 10, pages 109–122. Academic Press, New York and London.

[Medard et al., 2002] Medard, M., Barry, R. A., Finn, S.G., He, Wenbo, and Lumetta, S.S (2002). Generalized loop-back recovery in optical mesh networks. *IEEE/ACM Trans. Networking*, 10(1):153–164.

[Medard et al., 1999] Medard, M., Finn, S.G., Barry, R.A., and Gallager, R.G. (1999). Redundant trees for preplanned recovery in arbitrary vertex-redundant or edge-redundant graphs. *IEEE/ACM Trans. Networking*, 7(5):641–652.

[Metro Ethernet Forum whitepaper, 2002] Metro Ethernet Forum whitepaper (2002). Metro Ethernet networks – a technical overview.

[Miyao and Saito, 1998] Miyao, Y. and Saito, H. (1998). Optimal design and evaluation of survivable WDM transport networks. *IEEE J. Selected Areas in Communications*, 16(7):1190–1198.

[Modiano and Lin, 2001] Modiano, E. and Lin, P. J. (2001). Traffic grooming in WDM networks. *IEEE Communication Magazine*, 39:124–129.

[Modiano and Narula-Tam, 2002] Modiano, E. and Narula-Tam, A. (2002). Survivable lightpath routing: a new approach to the design of WDM-based networks. *IEEE J. Selected Areas in Communications*, 20(4):800–809.

[Mohan et al., 2001] Mohan, G., Murthy, C. S. Ram, and Somani, A. K. (2001). Efficient algorithms for routing dependable connections in WDM optical networks. *IEEE/ACM Trans. Networking*, 9(5):553–566.

[Mukherjee, 1997] Mukherjee, B. (1997). *Optical Communication Networks*. McGraw-Hill, New York.

[Mukherjee, 2000] Mukherjee, B. (2000). WDM optical communication networks: Progress and challenges. *IEEE J. Selected Areas in Communications*, 18(10):1810–1824.

[Mukherjee et al., 1996] Mukherjee, B., Ramamurthy, S., Banerjee, D., and Mukherjee, A. (1996). Some principles for designing a wide-area optical network. *IEEE/ACM Trans. Networking*, 14:684–696.

[Nucci et al., 2002] Nucci, A., Taft, N., Thiran, P., Zang, H., and Diot, C. (2002). Increasing the link utilization in IP-over-WDM networks. In *Proc. OptiComm*.

[Optical Internetworking Forum (OIF), 2004] Optical Internetworking Forum (OIF) (2004). http://www.oiforum.com.

[Ou et al., 2002] Ou, C., Zang, H., and Mukherjee, B. (2002). Sub-path protection for scalability and fast recovery in optical WDM mesh networks. In *Proc. OFC*, page ThO6.

[Ou et al., avis] Ou, C., Zang, H., Singhal, N., Zhu, K., Sahasrabuddhe, L., MacDonald, R., and Mukherjee, B. (Tech. Report CSE-2002-32, Oct. 2002, Dept. of Computer Science, University of California, Davis.). Sub-path protection for scalability and fast recovery in optical WDM mesh networks.

[Ou et al., 2003] Ou, C., Zhang, J., Zang, H., Sahasrabuddhe, L., and Mukherjee, B. (2003). Near-optimal approaches for shared-path protection in WDM mesh networks. In *Proc. IEEE ICC*, pages 1320–1324.

[Ou et al., 2004] Ou, C., Zhang, J., Zang, H., Sahasrabuddhe, L., and Mukherjee, B. (2004). New and improved approaches for shared-path protection in WDM mesh networks. *IEEE/OSA J. Lightwave Technology*, 22:1223–1232.

[Qiao and Xu, 2002] Qiao, Chunming and Xu, Dahai (2002). Distributed partial information management (DPIM) schemes for survivable networks – Part I. In *Proc. IEEE INFOCOM*, pages 302–311.

[Ramamurthy et al., 2001] Ramamurthy, R., Bogdanowicz, Z., Samieian, S., Saha, D., Rajagopalan, B., Sengupta, S., Chaudhuri, S., and Bala, K. (2001). Capacity performance of dynamic provisioning in optical networks. *IEEE/OSA J. Lightwave Technology*, 19(1):40–48.

[Ramamurthy and Mukherjee, 2002] Ramamurthy, R. and Mukherjee, B. (2002). Fixed-alternate routing and wavelength conversion in wavelength-routed optical networks. *IEEE/ACM Trans. Networking*, 10(3):351–367.

[Ramamurthy and Mukherjee, 1999] Ramamurthy, S. and Mukherjee, B. (1999). Survivable WDM mesh networks, Part II – restoration. In *Proc. IEEE ICC*, pages 2023–2030.

[Ramamurthy et al., 2003] Ramamurthy, S., Sahasrabuddhe, L., and Mukherjee, B. (2003). Survivable WDM mesh networks. *IEEE/OSA J. Lightwave Technology*, 21(4):870–883.

[Ramanathan, 1996] Ramanathan, S. (1996). Multicast tree generation in networks with asymmetric links. *IEEE/ACM Trans. Networking*, 4(4):558–568.

[Ramaswami and Sivarajan, 1995] Ramaswami, R. and Sivarajan, K. N. (1995). Routing and wavelength assignment in all-optical networks. *IEEE/ACM Trans. Networking*, 3(5):489–500.

[Ramaswami and Sivarajan, 1998] Ramaswami, Rajiv and Sivarajan, Kumar N. (1998). *Optical Networks: A Practical Perspective.* San Francisco: Morgan Kaufmann Publishers.

[Robertson and Seymour, 1995] Robertson, N. and Seymour, P. D. (1995). Graph minors XIII: The disjoint paths problem. *J. Combinatorial Theory Ser. B*, 63:65–110.

[Sahasrabuddhe et al., 2002] Sahasrabuddhe, L., Ramamurthy, S., and Mukherjee, B. (2002). Fault tolerance in IP-over-WDM networking: WDM protection vs. IP restoration. *IEEE J. Selected Areas in Communications*, 20(1):21–33.

[Sahin and Azizoglu, 2001] Sahin, Gökhan and Azizoglu, Murat (2001). Optical layer survivability for single and multiple service classes. *J. High Speed Networks*, 10(2):91–108.

[Sen et al., 2001] Sen, Arunabha, Shen, Bao Hong, and Bandyopadhyay, Subir (2001). Survivability of lightwave networks–path lengths in WDM protection scheme. *J. High Speed Networks*, 10(4):303–315.

[Simmons et al., 1999] Simmons, J. M., Goldstein, E. L., and Saleh, A. A. M. (1999). Quantifying the benefit of wavelength add-drop in WDM rings with distance-independent and dependent traffic. *IEEE/OSA J. Lightwave Technology*, 17:48–57.

[Sridharan et al., 2002] Sridharan, M., Srinivasan, R., and Somani, A.K. (2002). On improving partial information routing with segmented path protection. In *International Conference on Parallel Processing Workshops*, pages 193–198.

[Stanley, 2003] Stanley, S. (Lightreading report, March 2003). Making SONET ethernet-friendly.

[Stern and Bala, 1999] Stern, Thomas E. and Bala, Krishna (1999). *Multiwavelength Optical Networks: A Layered Approach*. Reading, Mass.: Addison-Wesley.

[Strand et al., 2001] Strand, J., Chiu, A.L., and Tkach, R. (2001). Issues for routing in the optical layer. *IEEE Communication Magazine*, 39(2):81–87.

[Strand and Xue, 2001] Strand, J. and Xue, Y. (OIF2001.046, Jan. 2001). Routing for optical networks with multiple routing domains.

[Su and Su, 2001a] Su, C. and Su, X. (2001a). Protection path routing on WDM networks. In *Proc. OFC*, volume 2, pages TuO2–T1–3.

[Su and Su, 2001b] Su, X. and Su, C. (2001b). An online distributed protection algorithm in WDM networks. In *Proc. IEEE ICC*, volume 5, pages 1571–1575.

[Suurballe and Tarjan, 1984] Suurballe, J. W. and Tarjan, R. E. (1984). A quick method for finding shortest pairs of disjoint paths. *Networks*, (14):325–336.

[Thiagarajan and Somani, 2001a] Thiagarajan, S. and Somani, Arun K. (2001a). Capacity fairness of WDM networks with grooming capabilities. *SPIE Optical Networks Magazine*, 2:24–32.

[Thiagarajan and Somani, 2001b] Thiagarajan, S. and Somani, Arun K. (2001b). Traffic grooming for survivable WDM mesh networks. In *Proc. OptiComm*, pages 54–65.

[Todimala and Ramamurthy, 2002] Todimala, Ajay and Ramamurthy, Byrav (2002). A dynamic partitioning sub-path protection routing technique in WDM mesh networks. In *15th International Conference on Computer Communication*.

[Van Caenegem et al., 1998] Van Caenegem, B., Van Parys, W., De Turck, F., and Demeester, P.M. (1998). Dimensioning of survivable WDM networks. *IEEE J. Selected Areas in Communications*, 16(7):1146–1157.

[Wang et al., 2004] Wang, D., Li, G., Yates, J., and Kalmanek, C. (2004). Efficient segment-by-segment restoration. In *Proc. OFC*.

[Wang et al., 2002a] Wang, D., Strand, J., Yates, J., Kalmanek, C., Li, G., and Greenberg, A. (2002a). OSPF for routing information exchange across metro/core optical networks. *SPIE Optical Networks Magazine*, 3(5):34–43.

[Wang et al., 2001] Wang, J., Cho, W., Vemuri, V. R., and Mukherjee, B. (2001). Improved approaches for cost-effective traffic grooming in WDM ring networks: ILP formulations and single-hop and multiphop connections. *IEEE/OSA J. Lightwave Technology*, 20(1):122–133.

[Wang and Mukherjee, 2002] Wang, J. and Mukherjee, B. (2002). Interconnected WDM ring networks: strategies for interconnection and traffic grooming. *SPIE Optical Networks Magazine*, 3(5):10–20.

[Wang et al., 2002b] Wang, J., Sahasrabuddhe, L., and Mukherjee, B. (2002b). Fault monitoring and restoration in optical WDM networks. In *Proc., National Fiber Optic Engineers Conference*, pages 262–269.

[Wang et al., 2002c] Wang, J., Sahasrabuddhe, L., and Mukherjee, B. (2002c). Path vs. subpath vs. link restoration for fault management in IP-over-WDM networks: Performance comparisons using GMPLS control signaling. *IEEE Communication Magazine*, 40(11):2–9.

[Wen et al., 2002] Wen, W., Mukherjee, B., and Yoo, S. J. B. (2002). QoS based protection in MPLS controlled WDM mesh networks. *Photonic Network Communications*, 4:297–320.

[Xin et al., 2001a] Xin, C., Ye, Y., Dixit, S., and Qiao, C. (2001a). A joint lightpath routing approach in survivable optical networks. In *Proc. SPIE Asia-Pacific Optical and Wireless Communications*, pages 139–146.

[Xin et al., 2001b] Xin, Chunsheng, Ye, Yinghua, Dixit, S., and Qiao, Chunming (2001b). A joint working and protection path selection approach in WDM optical networks. In *Proc. IEEE Globecom*, volume 4, pages 2165–2168.

[Xiong et al., 2003] Xiong, Y., Xu, D., and Qiao, C. (2003). Achieving fast and bandwidth-efficient shared-path protection. *IEEE/OSA J. Lightwave Technology*, 21(2):365–371.

[Xu et al., 2003a] Xu, D., Xiong, Y., and Qiao, C. (2003a). Novel algorithms for shared segment protection. *IEEE J. Selected Areas in Communications*, 21(8):1320–1331.

[Xu et al., 2003b] Xu, Dahai, Xiong, Yizhi, Qiao, Chunming, and Li, Guangzhi (2003b). Trap avoidance and protection schemes in networks with shared risk link groups. *IEEE/OSA J. Lightwave Technology*, 21(11):2683–2693.

[Yen, 1971] Yen, J. Y. (1971). Finding the K shortest loopless paths in a network. *Management Science*, (11):712–716.

[Zang et al., 2000] Zang, H., Jue, J., and Mukherjee, B. (2000). A review of routing and wavelength assignment approaches for wavelength-routed optical WDM networks. *SPIE Optical Networks Magazine*, 1(1):47–60.

[Zang et al., 2003] Zang, H., Ou, C., and Mukherjee, B. (2003). Path-protection routing and wavelength-assignment (RWA) in WDM mesh networks under duct-layer constraints. *IEEE/ACM Trans. Networking*, 11:248–258.

[Zhang et al., 2003] Zhang, J., Zhu, K., Sahasrabuddhe, L., Yoo, S. J. Ben, and Mukherjee, B. (2003). On the study of routing and wavelength-assignment approaches for survivable wavelength-routed WDM mesh networks. *SPIE Optical Networks Magazine*, 4:16–27.

[Zhang and Qiao, 2000] Zhang, X. and Qiao, C. (2000). An effective and comprehensive approach for traffic grooming and wavelength assignment in SONET/WDM rings. *IEEE/ACM Trans. Networking*, 8:608–617.

[Zhu et al., 2002] Zhu, H., Zang, H., Zhu, K., and Mukherjee, B. (2002). Dynamic traffic grooming in WDM mesh networks using a novel graph model. In *Proc. IEEE Globecom*, pages 2681–2685.

[Zhu et al., 2003a] Zhu, H., Zang, H., Zhu, K., and Mukherjee, B. (2003a). A novel, generic graph model for traffic grooming in heterogeneous WDM mesh networks. *IEEE/ACM Trans. Networking*, 11(2):285–299.

[Zhu and Mukherjee, 2002a] Zhu, K. and Mukherjee, B. (2002a). On-line approaches for provisioning connections of different bandwidth granularities in WDM mesh networks. In *Proc. OFC*, page ThW5.

[Zhu and Mukherjee, 2002b] Zhu, K. and Mukherjee, B. (2002b). Traffic grooming in an optical WDM mesh network. *IEEE J. Selected Areas in Communications*, 20(1):122–133.

[Zhu and Mukherjee, 2003] Zhu, K. and Mukherjee, B. (2003). A review of traffic grooming in WDM optical networks: architectures and challenges. *SPIE Optical Networks Magazine*, 4(2):55–64.

[Zhu et al., 2003b] Zhu, K., Zhu, H., and Mukherjee, B. (2003b). Traffic engineering in multi-granularity, heterogeneous, WDM optical mesh networks through dynamic traffic grooming. *IEEE Network*, 7:8–15.

[Zhu et al., 2003c] Zhu, Keyao, Zang, Hui, and Mukherjee, Biswanath (2003c). Exploiting the benefit of virtual concatenation technique to the optical transport networks. In *Proc. OFC*, pages 363–364.

Index

MIX
Papier aus verantwortungsvollen Quellen
Paper from responsible sources
FSC® C105338

If you have any concerns about our products,
you can contact us on
ProductSafety@springernature.com

In case Publisher is established outside the EU,
the EU authorized representative is:
Springer Nature Customer Service Center GmbH
Europaplatz 3, 69115 Heidelberg, Germany

Printed by Libri Plureos GmbH
in Hamburg, Germany